CW01239312

CROYDONOPOLIS

CROYDONOPOLIS

A JOURNEY TO THE GREATEST CITY THAT NEVER WAS

Will Noble

SAFE HAVEN

First published 2024 by Safe Haven Books Ltd
12 Chinnocks Wharf
14 Narrow Street
London E14 8DJ
www.safehavenbooks.co.uk

Copyright © Will Noble 2024

The moral right of Will Noble to be identified as the Author of this Work has been asserted by him in accordance with the Copyrights, Designs and Patents Act, 1988.

All Rights Reserved.
No part of this book may reproduced or utilised in any form or by any means, electronic or mechanical, including photocopying, recording or by any information storage and retrieval system, without permission in writing from Safe Haven Books Ltd.

Every effort has been made to contact the copyright holders of material in this book. Where an omission has occurred, the publisher will gladly include acknowledgement in any future edition.

A catalogue record for this book
is available from the British Library.

ISBN 978 1 8384051 9 9

1 3 5 7 9 10 8 6 4 2

Typeset in Caslon, HWT Slab and
Trade Gothic by M Rules Ltd

Printed and bound in the UK by
Bell & Bain, Glasgow

For Jojo. Croydon wouldn't be home without you.

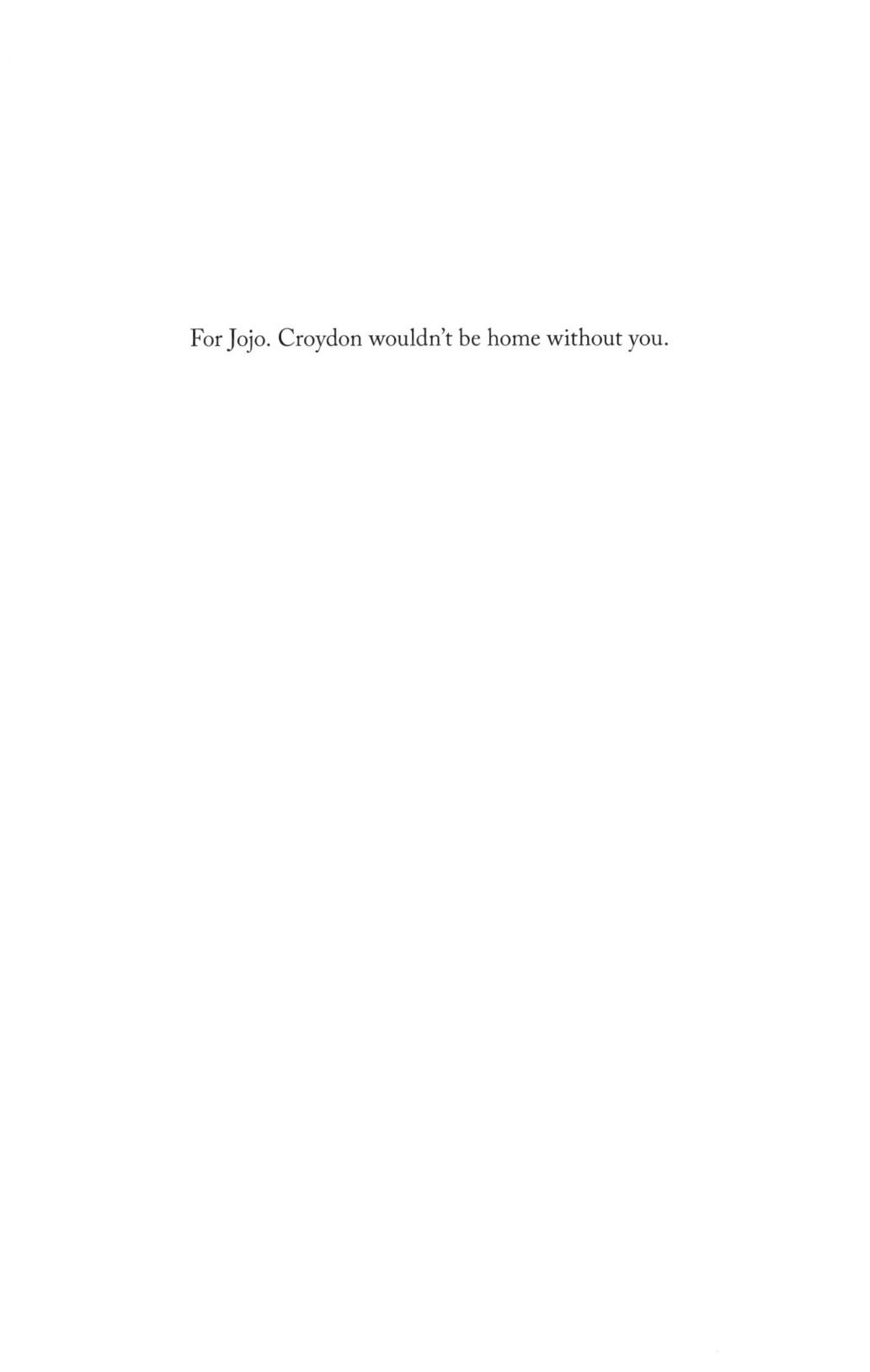

Ugliness is in a way superior to beauty because it lasts.

Serge Gainsbourg

You want the Moon? Just say the word and I'll throw a lasso around it and pull it down.

George Bailey, *It's a Wonderful Life*

CONTENTS

	Introduction	1
1	FOUNDING FATHER	9
2	SOVEREIGN POWER	19
3	THE WHITGIFT THAT KEEPS ON GIVING	25
4	THE IDEAL SITUATION	35
5	'NARNIA IN URBAN GREEN'	49
6	WELL TRAINED	63
7	RETAIL MAGNET	73
8	TAKING WING	85
9	CENTRE OF THE UNIVERSE	105
10	FALL FROM GRACE	117
11	CONCRETE RENAISSANCE	125
12	SKY-HIGH SCI-FI	139
13	THE SEVEN HILLS OF CROYDON	151
14	CROLLYWOOD	161

15	HERE COMES RUBBLE	171
16	'I COULD BLOODY PLAY THERE ALL NIGHT!'	183
17	SMASH IT UP	195
18	TRUE BRIT	207
19	COSTA DEL CROYDON	217
20	THE TOWN CENTRE CANNOT HOLD	227
21	CROYDONOPOLIS	243
	Acknowledgements	259
	Croydonography	263
	Index	273

Introduction

COME, FRIENDLY BOMBS, AND FALL ON CROYDON

Here's something that'll make you laugh.

Croydon!

'Less of a place, more of a punchline', according to the Croydon-born Sue Perkins.

'Croydon?' sneered Kenneth Williams. 'Sounds like an *illness*.'

'Croydon!' sang Harry Hill in an improbable power ballad:

> City of a thousand dreams
> I love Croydon!
> Has a very strong branch of Nando's.

Croydon: home to the toadying Captain Kevin Darling in *Blackadder Goes Forth*.

Croydon: where the Channel 4 sitcom *Peep Show*, in which the two perma-losers Mark Corrigan and Jeremy Usborne fritter away their lives in a pebbledash purgatory, is set. 'Is there anywhere nice around ... um ... here?' asks a woman who's just regretfully talked herself into a date with Mark. 'No,' replies Mark, quick as a flash.

'Some instinct, or mere snobbery', writes Ian McEwan of Florence Ponting, the young musician in his novel *On Chesil Beach*, 'convinced her she could not live in or near Croydon'. In Robert Smith Surtees's 1830s picaresque capers *Jorrocks' Jaunts and Jollities*, the then-countrified denizens of Croydon are derided as 'chaw-bacons' who 'retain much of their pristine barbarity'. Fast-forward 150 years and David Lodge's novel *Paradise News* introduces us to the middle-aged Lilian from Croydon, a 'roly-poly figure stuffed into matching electric blue stretch-pants and jumper'. In Alan Judd's 2014 thriller *Inside Enemy*, the spy Charles Thoroughgood is happy to be called back into the fold but horrified to find MI6's HQ has been shifted to a CR0 postcode: 'Already it felt good to be back in harness, to be wanted again. Except for Croydon.'

Even Hollywood sticks the knife in. 'They say his Lear was the toast of Croydon,' smarms Guy Pearce as Aldrich Killian in *Iron Man 3*, 'wherever that is.' In the violent 2019 comedy *The Gentlemen* Colin Farrell's character Coach is told to look for the boxer he's after in the 'posh part of Croydon'. 'There *is* no posh part of Croydon!' snaps Coach. At least the scenes for Danny Boyle's zombie film *28 Days Later* that were shot around the Croydon Flyover were left on the cutting room floor as 'not quite apocalyptic enough'.

'I married a man of the Croydon class,' goes Anna Wickham's poem 'Nervous Prostration',

> When I was twenty-two.
> And I vex him, and he bores me
> Till we don't know what to do!
> It isn't good form in the Croydon class
> To say you love your wife,

So I spend my days with the tradesmen's books
And pray for the end of life.

Henry VIII was one of many reluctant visitors to Croydon, claiming it made him feel queasy. To Sir Francis Bacon it was 'an obscure and darke place'. 'A Merry Christmas', the flamboyant racing tipster Prince Monolulu, sporting an ostrich feather headdress and tartan cummerbund, told the judge at Croydon Magistrates Court one December on his conditional discharge. 'I will never come back here again.' 'I'm certain that of all the places I've visited,' wrote Tom Chesshyre in *To Hull and Back*, a travelogue that sets out to visit the lesser-appreciated parts of Britain, including Hull, Port Talbot and somewhere called Hell Bay, 'I'm least likely to return to Croydon.'

If you're to believe a 1995 article in the *Independent*, indeed, Croydon was the original target for John Betjeman in his best-known poem that begins, notoriously, 'Come, friendly bombs, and fall on Slough.' It was only spared the literary levelling, apparently, because Betjeman couldn't find a rhyme.

But what of the real-life Croydonians who had no choice but to be born and raised here? The great film director David Lean made his childhood in Croydon bearable by superimposing the films he'd just seen in the cinema onto the 'dreadful town' that was Croydon. 'As far as growing up in Croydon went, it wasn't a time I look back on longingly,' the supermodel Kate Moss told *Reader's Digest* once, 'certainly not so much as the place itself. I was keen to get away.'

'There is no getting away from the fact,' Gavin Barwell told Parliament in 2010, 'that Croydon has an image problem.' He had just been elected to represent the town as its MP. Google 'Croydon', and the search helpfully suggests questions like 'Is there a nice part of Croydon?', 'How rough is Croydon?' and 'How many stabbings in Croydon this year?' On Instagram, accounts like Croydon Extra depict a lawless, topsy-turvy town in which residents ride down the main road on a skateboard pulled by dogs, or pick a fight with a tram. *Lion King* memes show Mufasa and Simba gazing out over a golden horizon. 'Look, Simba, everything the light touches is Surrey,' says Mufasa to his starry-eyed cub.

'Wow,' replies Simba. 'But what about that shadowy place?'

'That is Croydon. You must never go there, Simba.'

'It was my nemesis, I hated Croydon with a real vengeance,' declared David Bowie, who studied at Croydon Art College, to *Q* magazine in 1999. 'I think it's the most derogatory thing I can say about somebody or something: "God, it's so fucking *Croydon*!"'

As the 1970s countercultural 'zine *Suburban Press* warned in no uncertain terms: 'CROYDON IS DEATH TOWN.'

'It's either going to be like this . . .' cackled a stall owner on Surrey Street Market from behind a gleaming mountain of cherries, the gap between her fingers indicating a wafer-thin pamphlet, 'or this' – and using both hands she enlarged the invisible tome to biblical girth.

All I'd done was mention I was writing a book about Croydon. But I was getting used to this kind of reaction. As I was explaining to a driver at the Croydon tram depot what one of the chapters was about, he interrupted with a deadpan '—And then eight chapters on anti-social behaviour?' By chance, I ran into Sue Perkins and told her about the book. 'Oh god,' she said.

Cards on the table: I've not always had constructive things to say about Croydon either. In 2018, I wrote a piece for *Londonist* which fingered Croydon High Street as the home of 'London's Dullest Plaque', commemorating 'the widening of Croydon High Street from a width of 29 feet to its present width of 50 feet'.

So why the relentless bad press? What *is* so dreadful about Croydon? How did this town of 200,000 people in London's southernmost borough wind up as a trope for everything that is banal, ugly and naff?

Granted, it's not the prettiest town on the map. Reshaped by a maniacal outbreak of reconstruction between the late 1950s and early 1970s, the core of Croydon doesn't contain one or two white elephants, but a whole safari park. An architectural wag once commented that Croydon 'isn't even interestingly bad'. Others have denounced it as an 'English Alphaville' or 'New York in Eastern Europe'. In 1971, the *London Evening News*'s Rex Grizell lamented 'a place of hard, flat surfaces, corners, reflections and cold winds'. In 1996, the *Independent* called it 'the high-rise epicentre of all things suburban, bureaucratic and dull . . . the spiritual home of shopping malls . . . of superstores, DIY centres and of polyester-nylon suits.' 'You think of Croydon and you sigh rather deeply,' mused the architectural historian Colin Amery. Croydon's adversary, David Bowie, was even more succinct: it was a 'complete concrete hell'.

In the eighteenth century Croydon was riddled with smugglers and murderous highwaymen; by the nineteenth it had established its own police force to deal with troublemakers, the snag being that officers were often so inebriated they could barely stand up, let alone arrest anyone. By the late 1970s headlines were warning of roving gangs of sledgehammer-wielding 'terrorists' – some as young as *four* – smashing windows, posting excrement through letterboxes and setting flats alight.

And yet when Croydon tries to better itself people don't like that either. In 1996, bold plans were unveiled to construct the 'Tower of Light', a mast of steel and billowing fabric sails, in the centre of town. The idea was to buoy the spirits; give Croydon a lambent totem pole to gaze up at and be proud of. 'It's a tacky, tasteless, *Star-Trek*ian folly that will boldly go where no man will want to go again,' complained one Labour councillor. 'If this is built', added another, 'Croydon will be the butt of jokes once again.' 'It's hideous,' said the *Evening Standard*'s art critic, Brian Sewell. 'If they had invested in decent buildings in the first place, they wouldn't be having to jump through hoops now.' The Tower of Light never saw the light of day.

Croydon's most notorious attempt to scale the ladder of respectability has been its string of failed applications for city status. These began in the 1950s, and have been consistently rebuffed ever since, like a contestant on *Britain's Got Talent* ploughing on with an increasingly excruciating tap-dancing routine even as Simon Cowell smashes his forehead on the buzzer.

Indeed, there are reasons to dislike Croydon you probably weren't even aware of. On 21 April 1807, one of the last ever British slave ships to set sail from England was called the *Croydon*, owned by a slave trader, Alexander Caldcleugh, who lived in the town. Addiscombe, just outside central Croydon, was the home of the East India Company Military Seminary, a college where young men were primed to invade, plunder and lord it abroad. The roads stippled around this part of Croydon are still named after East India Company officers. 'I lived in Addiscombe and I absolutely hated walking past that,' the Bishop of Croydon, Rosemarie Mallet, told me.

Even Croydon itself seems unsure of its own identity. Terms like 'dormitory town' and 'edge city' cast a halo of nebulous fuzziness around it. As the late architect and designer Charles Jencks once said, 'When you're there, there's no there *there*.' Half the people who live in Croydon will tell you it's South London; the other 50 per cent (often those of a certain generation, and invariably estate agents) will swear blind that it's in leafy Surrey. In 2019, police announced they were looking for a gunman with a 'Croydon accent', only for locals to assure them that *there is no such thing*.

Whatever it is that people know Croydon for – and they'll invariably know it for *something* – it's unlikely to be positive. It'll be the 2011 riots, civic corruption, a *Time Out* article dubbing it 'one of the most depressing neighbourhoods in the UK', the tragic tram crash of 2016, lumpen 1960s office blocks, unenviable 'knife crime capital' titles, the pejorative term 'Croydon facelift', cat decapitations, *Private Eye*'s 'Ultimate Rotten Borough' award, or an unhealthy obsession with multi-storey car parks. Croydon, received wisdom tells us, can't have nice things. And while we're talking punchlines, now seems as good a time as any to mention that Croydon's official motto is *'Let us strive for perfection'*.

In 2022 I moved to Croydon. When I bumped into the editor of the *Inside Croydon* website on the staircase of the Fairfield Halls and told him, he grinned: 'Didn't you *read* any of my articles?' Although part of the freshet of younger Londoners pushed ever southwards by the (un)affordability of housing, I already had a soft spot for Croydon, however. My first trip a number of years earlier had included a visit to the Museum of Croydon, which piqued an interest in the local history. A return visit for a *Londonist* article – on the same day, as it happened, that Croydon's anti-ambassador David Bowie left

Ronnie Corbett, Samuel Coleridge-Taylor and Peggy Ashcroft form the unlikeliest of sculptural supergroups on Charles Street in central Croydon. © *Will Noble*

CROYDONOPOLIS

Planet Earth for good – took me to Croydon Airport, to be shown around by Ian Walker, a charming commercial airline pilot with an encyclopaedic knowledge of this halcyon chapter in the town's past. My love for Croydon swelled that day.

And now, a fully-fledged Croydonian, one of the first things I noticed on my wanderings were three steel cut-out statues on Charles Street. Who were they? One, I discovered, was the Oscar-winning actor Peggy Ashcroft; another was the avuncular comedian Ronnie Corbett (someone once wryly attached four candles to his one); the third was the Romantic composer Samuel Coleridge-Taylor. The world's unlikeliest supergroup: here was confirmation that something was bubbling beneath the bleak Croydon stereotype. It had to be hauled to the surface and laid out for all to see.

I started knocking around the shopping centres, riding trams into the wooded outskirts, going on guided tours, rummaging through stacks of green boxes at the local archives, watching shows at the Fairfield Halls and Oval Tavern. One by one, tales emerged of archbishops, bubble cars, starchitects and rock stars. Croydon had its own Poet Laureate too – and she hadn't mentioned 'friendly bombs' once. Even that tedious plaque, it turned out, had a good reason for being there.

There wasn't, I discovered, just one golden era of Croydon, but a whole cornucopia. And it wasn't simply the case that a book like this *could* be written, but that it *should*. If nothing else, to address one of the laziest, most inexcusable cases of long-running slander ever. Croydon, you see, is a very special place with a fascinating story to tell, and a moral for us all.

If you're waiting for the '*ba-dum – TISH!*', I'm afraid there isn't one.

Will Noble
Croydon
2024

1
FOUNDING FATHER

The sweet tang of katsu curry drifts through the Whitgift Centre, but shoppers these days are thin on the ground. Here and there the odd pram is steered around the obstacle course of bright yellow buckets marked 'Caution, Slip Hazard', rainwater drooling into them from the ceiling above. There are so many of these unsightly receptacles one YouTube vlogger recently quipped that the Whitgift looked less like a shopping centre, more like a level on *Crash Bandicoot*. An eeriness pervades. As you ride up the escalators with varying degrees of death rattle, you almost expect to see a pack of flesh-eating zombies pawing at the windows of WH Smith. 'First and last visit, horrible' and 'Dump', report terse TripAdvisor reviews. With its moribund mezzanines of half-vacated outlets and post-nasal ceiling, the Whitgift is somewhere that has seen better days. There is a metaphor here somewhere.

Whitgift. The more you look, the more you see the name scattered about Croydon. The Whitgift Centre. A Whitgift News corner shop. Whitgift care homes. There are Whitgift schools, the John Whitgift Community Cube at Fairfield Halls, and the Whitgift multi-storey car park. Vans with 'Whitgift Hire' blazoned on the side whizz along Whitgift Street. The word

'Whitgift' seems almost as ubiquitous as 'Croydon' itself. And the owner is a man who, a good few hundred years ago, would make sure that Croydon was never going to be just any old town.

Croydon has always been on the way to somewhere else – something that's proved both a curse and a blessing. In the case of the Archbishops of Canterbury, it was more of the latter.

In the days when Croydon's Old Town – the low-lying swathe of land just to the west of the modern town centre – brimmed with bucolic goodness, the journey between Canterbury and Lambeth, even for archbishops en route from cathedral to palace, was a staggered road trip made on horseback. The archbishops had a string of residences – a form of ye olde motels – peppered along the way at 10-to-15-mile intervals, and Croydon was one such pitstop.

It was here, then, that the Diocese of Canterbury decided to build a manor house: Croydon Palace (or Croydon Manor, or Croydon House, as it was called early on). Sometimes archbishops would only stay the night before hitting the road again, but at others the Elysian setting of Croydon would persuade them to tarry longer, raid the wine cellar and get stuck into a few lampreys.

Croydon Palace cuts an unlikely figure in the modern-day town. Turn off Church Street just as you get to Lidl, walk a couple of hundred yards, and there it is. This is not a palace in the Buckingham or Blenheim sense; more a higgledy-piggledy clutch of buildings: red herringbone here, half-timbered beams there. Most prominent of all is the rust-red triangle thrusting up into the heavens like a wedge of aged Red Leicester.

It houses the Great Hall – built by Archbishop John Stafford in the mid-fifteenth century. While Croydon Palace is not awash with Japanese tourists, it does occasionally open for public tours, giving you the chance to see one of the country's earliest uses of brick, then step inside the Great Hall and marvel at its hearty timber roof, probably cut from the Great North Wood, which once swept from Deptford in the north to the fringes of Croydon in

FOUNDING FATHER

The Whitgift Shopping Centre is now a husk of its former self, and would make a pretty good set for a George A. Romero-style zombie flick. © *Will Noble*

the south. Tucking into tea and shortbread, you can gape upwards at the heraldic angels, while listening to guides tell how Henry III once ordered 252 gallons of wine to Croydon Palace, for what must've been one hell of a house party. The angels on one side of the hall bear royal coats of arms; on the other, the coats of arms belong to bishops. Successive archbishops painted over the shields of their predecessors, and Archbishop Herring is among those who had the parting shot. His coat of arms features a shoal of fish.

Archbishop Lanfranc was the first Archbishop of Canterbury known to have owned land in Croydon (it's recorded in the Domesday Book of 1086) and, as successive archbishops trickled in, the palace swelled in size with various embellishments and extensions. The archbishops began to leave their mark on Croydon itself: gouging crosses into trees and stones to demarcate areas of special privilege, and fencing off swathes of land for hunting. In 1276 Robert Kilwardby became one of Croydon's early endorsers when he acquired a charter to hold a market – one that happily bustles today, and still only accepts coins of the realm. By this time, says Oliver Harris in *The Archbishops' Town*, 'Croydon already had something of the physical character of a town.' But not all archbishops had a kinship with the locals. Edmund Grindal has the dubious honour of being Croydon's original pantomime villain. Arriving in 1576, having graduated from Archbishop of York to Archbishop of Canterbury, Grindal soon proved himself an incorrigible sourpuss by banning Whitsun ales and Morris dancing. But his reputation with Croydonians would be cemented by a dispute over smoke.

Roosting on the southern edge of the Great North Wood, Croydon was home to throngs of colliers who plied their trade by turning felled wood into charcoal. In a time before pit coal and sea coal, Croydon produced the bulk of charcoal that was hurled into the City of London's fires and furnaces. This thriving industry had its side effects. When the charcoal kilns were dampened (a vital part of the charcoal-making process), epic plumes of black fug billowed across the land. Croydon was the 'Big Smoke' long before London earned that moniker. When the wind blew in a southerly direction, the miasma was carried right across Grindal's idyllic estate at Croydon. This didn't sit well with the ornery archbishop at the best of times, but on one occasion the smoke was particularly troubling. 'What means this smother?' Grindal roared at his chamberlain, according to *The Phoenix Suburb* – 'Is good Croydon town ablaze?!'

A legal tussle ensued between Grindal and the colliers, who were represented by a Master Grimes. Remarkably, Archbishop Grindal lost the case on the grounds that *God* decides which way the wind blows. A sore loser, the shellacked archbishop later refused Grimes permission to be buried with his family in Croydon churchyard. But the plucky collier came up trumps in the

FOUNDING FATHER

Many Archbishops of Canterbury used Croydon Palace as a picturesque pied-à-terre, not to mention Elizabeth I when going to watch the horse racing at nearby Duppas Hill. © *Will Noble*

long run, becoming a literary folk hero in various texts, including the 1662 play *Grim, the Collier of Croyden*.

Following the Grindal debacle, Croydonians were desperate for an archbishop who'd see things their way, and in 1583 they got him. He arrived in the form of a shortish man with 'coal-black eyes and hair, wearing a beard rather long and thick'. His name was John Whitgift.

Born in Grimsby (then called Great Grimsby) around 1530, Whitgift was the formidable son of a sea merchant, who rose briskly through the ecclesiastical ranks, becoming Dean of Lincoln, then Bishop of Worcester. Whitgift was ambitious, politically astute and, when made Archbishop of Canterbury by Elizabeth I, soon reversed Grindal's kibosh on beer and general merriment. He didn't seem to have a beef with the colliers either. Whitgift was a keeper.

The feeling was mutual, because Whitgift soon fell in love with Croydon. He had a soft spot for his new palatial pied-à-terre, its orchards heaving with fruit trees, and the River Wandle, bristling with trout,[*] babbling past the windows. 'I give ... my pleasant open ayre and fragrant smels,' gushes the character of Summer in a play written at the time, 'to Croyden, and the ground abutting round.' On the palace's doorstep lay bosky woodlands, ready and waiting for whenever Whitgift felt like blowing off steam by emptying arrows into a stag or two. The palace was close enough to London for a handy commute, but far enough away that Whitgift could keep his distance from those on the Privy Council who gave him the stink-eye.

By the time Whitgift arrived in Croydon, the palace was more noteworthy than ever. A petulant Henry VIII had snatched many palatial estates back off the Church in the mid-1500s during the dissolution of the monasteries. Croydon, though, was let off the hook (maybe because Henry couldn't stand the place, once moaning, 'It standeth lowe, and is reumatike, where I could never be without sycknes'). Henry hadn't been entirely wrong. Though parts of Croydon were picturesque in Whitgift's day, it was far from utopian. A poem by Patrick Hannay from 1622 sneers at a town 'cloth'd in blacke. In a low bottome sinke of all these hills; And is receipt of all the durtie wracke', going on to bemoan its 'unpav'd lanes with muddie mire it fills', before concluding, 'You may well smell, but never see your way.' Whitgift would've known the centre of Croydon as a muddle of weatherboarded hovels dangling over the streets, with streams of blood and offal oozing down Butcher's Row (also known by the grim sobriquet the 'Fleshmarket'), where drovers brought their cattle to be slaughtered.

---

[*] In the early 1990s, architects proposed paying homage to Croydon's trout-heavy history by turning all the street lamps blue, making the roads look like streams, and therefore giving traffic the appearance of swimming fish. It never happened.

Most working Croydonians were farmers (oatmeal was a major Croydon export), blacksmiths, farriers, or otherwise colliers coming home from a hard day's charcoal-burning, all phlegmy coughs and be-sooted faces, aka 'Croydon Complexyone'. Flooding was common in the Old Town, and it wasn't unusual to find yourself ankle-deep in feculent water. Poverty, disease and hard times were on Whitgift's doorstep, and as he spent an increasing amount of time in Croydon he decided something must be done about it. And so it wasn't Grimsby or Canterbury that Whitgift shared his annual salary of £2,215 with, but this up-and-coming, if rough around the edges, Surrey market town. Payday had arrived for Croydon.

The Whitgift Almshouses perch on the top of Crown Hill, where trams rattle down the steep incline of Church Street, their trilling bells shooing Deliveroo cyclists out of the way. With their leaded mullion windows and bonny chimney stacks, the almshouses might look less socially awkward in, say, Chester or Cambridge. But here they are, smack-dab in the centre of Croydon, just across from Primark. In the 1830s, the Croydonian writer Thomas Frost recalled peeking through the almshouses' archway to see 'bent old men and women sunning themselves in the prim little court-yard'.

Residents still live here now, and the house rules haven't changed that much. To qualify as a resident you must be 60 or over, a practising Church of Englander, of 'meagre means', and from Kent, Lambeth or Croydon. Above the almshouse entrance is the Latin motto *Qui dat Pauperi non indigebit* ('He who gives to the poor shall not lack'). The quadrangle hiding just beyond this is a curious sixteenth-century bubble. Sounds of the mic-ed up street preachers and buskers on the North End are muffled by the Elizabethan brick. Well-kempt lawns are bordered with bright splashes of daffodils and blazing firethorns. And in the middle of this picture of tranquillity stands the bronze likeness of Whitgift: flowing robes, crisp ruff, scholarly hat, a sharp beard – and a stack of untitled tomes, which might as well be his 'to do' list for Croydon.

In 1596, Whitgift purchased the Chequer Inn on this spot, had it knocked down and built the almshouses (originally the Hospital of The Holy Trinity) in its place. Without knowing it, he'd set a precedent for tearing down buildings and starting from scratch – one Croydon has revelled in ever since.

Whitgift personally laid two cornerstones, a PR move that announced in no uncertain terms to the townsfolk, 'I did this, and you are very welcome.' The hospital was quickly followed by a small school and headmaster's house nearby. Whitgift's philanthropic legacy was being set, quite literally, in stone.

The almshouses were evidently Whitgift's baby, because he came here regularly, often dining with residents, and even staying overnight. 'He visited so often,' wrote the seventeenth-century biographer Izaak Walton, 'that he knew the names and dispositions, and was so truly humble that he called them brothers and sisters.'

But the construction of the almshouses also provided a glimpse into a darker side of the archbishop. Records show that on one occasion Whitgift kicked off at one of the builders for the inferior quality of his bricks. It must have been quite a tirade, because said builder 'burste into tears, saying, he was never the like served in anie work; he was ashamed of it, he could not excuse it, it was the wickedness and deceitfulnesse of the yearthe.' Whitgift did not suffer fools gladly. In another outburst, he called members of the clergy 'boys, babes, prinnocks, unlearned sots.' He was just getting warmed up.

Elizabethan England is often thought of as anti-Catholic, but as part of Elizabeth I's bid to balance things out, Whitgift was tasked with treating Puritans with, as the nineteenth-century historian Thomas Babington Macaulay put it, 'exceptional intolerance'. Macaulay also accused Whitgift of being 'a narrow-minded, mean, and tyrannical priest', while Whitgift's contemporaries painted a picture of a splenetic figure 'with a grimme and angrie countenance'. The Puritan Henry Barrowe, who was hounded by Whitgift, said, 'He is a monster, a miserable compound,' while Whitgift's

FOUNDING FATHER

John Whitgift was, in many ways, the founding father of Croydon, his legacy still reverberating around the town and borough today. *Alamy*

own aunt claimed he was 'a devil'. The Welsh Puritan John Penry must have had similar thoughts. When Penry complained that the Church of England hadn't translated enough bibles into Welsh, Whitgift blithely signed Penry's

17

The Whitgift Almshouses, built by John Whitgift at the end of the sixteenth century, are one of many unexpected slices of history to be discovered in Croydon's town centre. © *Will Noble*

death warrant, forbidding him to say goodbye to his four young daughters before his execution. 'John Whitgift was compassionate,' says Bill Wood, a curator and historian at the Whitgift Foundation, drily, 'except when he put people to death.' Wales has never forgiven this. In 2008 it demanded that the then-Archbishop of Canterbury Rowan Williams apologise for Penry's death. While Whitgift's name runs through modern-day Croydon as through a stick of rock, it is also spat on by an entire country.

No surprise, then, that at least one attempt was made on Whitgift's life, which was one of the reasons why, whenever he came to Croydon, he clattered into the streets accompanied by an entourage of 800 horsemen.

But someone else had his back too. Elizabeth I.

2
SOVEREIGN POWER

'She probably listened to him more than she did sometimes to her own court,' says Bill Wood – hence why the Privy Council looked on Whitgift with such suspicion. It's no exaggeration to say that Elizabeth was a Whitgift fangirl – she probably even whispered things to him she hadn't divulged to her spymaster, Francis Walsingham. You sense a whiff of monarch–archbishop flirtation. Elizabeth, who had a penchant for unmarried men, once joked that Whitgift 'hath a white gift indeed', and was so proud of her pun that in framed charters on the wall of the almshouses' audience chamber the archbishop's name is written in immaculate calligraphy as 'Whitegift'. In return for her support, Whitgift did much of the Queen's bidding, and ensured that plots, rebellions and other such inconveniences were kept to a minimum, at whatever grisly cost to the likes of Barrowe and Penry.

Elizabeth was also a big admirer of Croydon, visiting once, twice, sometimes three times a year. Keen to spend quality time with her compadre Whitgift, she often arranged her visits to coincide with the horse racing at nearby Duppas Hill, now at the far end of the Croydon Flyover. Elizabeth was a sucker for the gee-gees, and in 1585 attended twice in a month. The following year, the equivalent of a royal box was built at Duppas Hill for her: 'a newe frame with a flower in it for the Quenes Majestie, ye noblemen, and ladies to stande in, to see the race runne with horses nere Croydon.' So well-known was the Queen's predilection for the horses, it could have been the end of her. In *The Croydon Races*, Jim Beavis explains how two fanatically pro-Pope priests, George Beesley and Montford Scott, had it in for her, and one was seen at the racecourse brandishing a pistol. Aware of the danger, Elizabeth stayed away, and the two would-be assassins were eventually captured and executed. After this scare, Elizabeth never went to the horses at Croydon again.

Racing in the town, on the other hand, would go from strength to strength. In 1859 at what is now Park Hill, barely half a mile from East Croydon Station, a crowd of 15,000–20,000 amassed to watch Croydon's Grand Open Steeplechase handicap. Even if the scene often resembled Glastonbury after an especially torrential weekend, the views from Park Hill were stunning, taking in 'the glittering Crystal Palace', the 'clean new villas' of Beulah Hill and Westow, and 'the well-timbered Addington Hills, Croham Hurst, and the valley beneath'.

By the 1870s the jumps had moved to Croydon's new racecourse to the north-east of the town at Woodside, on the site of what is now Ashburton Park. With its own dedicated railway station, the course was second in importance only to Aintree. Crowds were in raptures over Woodside's 'sensation' water jump, a tricksy five-foot fence where horses and their riders would often collapse into the mud. At Woodside tram stop today you can still see the ramp once used to ferry thoroughbreds on and off the trains. But less than 20 years later the NIMBYs successfully opposed the renewal of Croydon racecourse's licence, its last steeplechase was won in winter 1890 by Arthur Nightingall on Old Malt, and the racing headed down to Gatwick – not the last time Gatwick would swipe one of Croydon's booming industries.

*

One of the boons of having archbishops stay at Croydon was that it often drew royalty into the town. Henrys III, IV, VI and VII stayed at Croydon, as did Edwards I, II and III and Mary. It's also thought that Henry VIII proposed to Katherine of Aragon here. In these relatively formative days, then, Croydon already had the glint of prestige about it. King James I of Scotland, on the other hand, wasn't so keen. In 1414 he was kept prisoner at Croydon Palace in the custody of Archbishop Arundel. Maybe it's here that James was inspired to write his poem 'The King's Quire', which begins, 'Bewailing in my chamber thus allone, Despeired of all joye and remedy . . .'[*]

Still, Elizabeth knew how to have a time of it when she was in Croydon. She insisted on having her own bed lugged around the country as part of her peripatetic courts, infamously sapping the finances of her hosts. Whitgift threw lavish banquets for her in the palace's Great Hall, and would've watched the queen dance in the Long Gallery upstairs – the room in which she made Christopher Hatton (of Hatton Garden fame) Lord Chancellor of England, between bouts of dancing galliards and pavanes. When Elizabeth came to Croydon in 1592 Whitgift had a new play commissioned for her, *Summer's Last Will and Testament*, which fawned over 'Eliza Englands beauteous Queene, on whom all seasons prosperously attend'. The queen was notorious for imposing herself on her courtiers, but the wealthy Whitgift seemed all too eager to please – no expense spared.

Could it be that a gnawing guilt over these palatial extravagances prompted John Whitgift's charitable spending sprees across Croydon? Bill Wood isn't convinced, although perhaps there was something else troubling

---

[*] The presence of royals in Croydon is sometimes to be taken with a pinch of salt. At a South Croydon Liberal Society fete in the summer of 1930, guests were astounded to be graced by a surprise visit from Prince and Princess Henry of Bosnia-Herzegovina, who enjoyed a fevered reception of cheering and hat-waving, before the prince made a short speech in 'captivating broken English', then mingled with the starstruck hoi polloi. It turned out that the pair were in fact a couple of audacious Bright Young Things – an RAF captain and a young actor – and the *Croydon Times* journalist who was carried away in all the excitement, even going as far as to doff his cap to the duo, must've felt pretty daft when the penny dropped.

Whitgift: it's rumoured he was in a gay relationship with the five times Vice-Chancellor of the University of Cambridge, Andrew Perne. Puritan satirists even dubbed him 'Perne's Boy'. There's nothing to prove or disprove this, so whether this was a story of forbidden love or a malicious rumour spread by Whitgift's enemies, we'll never know.

John Whitgift's foes probably did have a hand in wiping him from the history books, though. As Archbishop of Canterbury he became one of the most important people in the country. Few others had his clout. When Elizabeth I was on her deathbed, it was John Whitgift she called for – it's possible she even died with him by her side (a scene immortalised in a stained-glass window in Grimsby Minster). Certainly Whitgift was deemed important enough to be the Chief Mourner at her funeral, and followed that up by crowning James I King of England and Ireland. It's bizarre that nowadays no-one really knows who he is.

Over the years Croydon Palace lost its saintly sheen. In 1652 the Cromwellian William Brereton treated it less than reverentially. He's described as having 'terribly long teeth and a stomach, to turn the Archbishop's Chapel of Croydone into a Kitchen.' By 1758 the palace was 'in a low and unwholesome situation, incommodious and unfit to be a habitation of an Archbishop of Canterbury'. DIY not being their strong point, the archbishops moved out. As the Industrial Revolution cranked into life, a water wheel was stuck on the side of the Great Hall, and the palace became a calico factory, then a bleaching works and, the final insult, a laundry, with sheets hung to dry from its rafters.

By the time the Sisters of Mercy of the Church of England inherited the palace in 1887, it was mostly in ruins. But the Sisters lived up to their name, organising cake sales and donkey rides to raise funds and turn it into a school. They scavenged glass from other churches, and even recycled the Tudor font from St George's in Southwark (the same font Dickens used to christen Little

Dorrit). A few years back, the school received an email from St George's, subject line: 'Have you got our font?' Elizabeth I's old bedroom, affectionately called the 'QER' (Queen Elizabeth's Room), is now a classroom, with bullet points about Islam and Sikhism pinned on the walls. Unlikely juxtapositions (like the digital projector balanced on the top of an ancient coat of arms in the Great Hall) are everywhere today, because the building is – for the time being – the Old Palace of John Whitgift School for girls.

As for the archbishops, they'd found a new country retreat, just out of the centre of town, namely Addington Palace.

That's right: Croydon doesn't just have one palace, but two.

3

THE WHITGIFT THAT KEEPS ON GIVING

If the image of a man riding a shark down a staircase while repeatedly punching it in the back of the head sounds somehow familiar, chances are you know Addington Palace.

Built on the site of an old manor house in 1774, the Palladian-style pile is the epitome of English refinement, and where Archbishop Charles Manners-Sutton decided to move his Croydon summer residence to in 1807. The surrounding fairways of the golf course, laid out in the 1930s, have entertained Bob Hope and Ronnie Corbett, and are now peppered with Wilson-swinging retirees any given day of the week.

Like Croydon Palace, Addington's has a Great Hall; its centrepiece is a pinkish-red marble fireplace that looks as though it's been carved from a slab of raw beef. It's guarded by a brace of bronze hunting dogs; a nod to when Henry VIII had a hunting lodge here. It's said Henry used the lodge to court Anne Boleyn in secret (for somewhere he loathed so vehemently, Henry did a prodigious amount of courting in Croydon). In one of the reception rooms now, the Tudor king's hirsute countenance lifts its eyebrows at newlyweds as if to say, 'Well, don't ask *me* for advice.'

Weddings aside, Addington Palace gets its income nowadays from TV and movie work. In one advert filmed here, Kate Moss frolics about in her underwear before running off with a dishy bellboy. The palace's stately wood panels and townhouse-esque staircase have made it a popular stunt double for Downing Street: in Girls Aloud's *Jump* video (a tie-in for the romcom *Love Actually*), the feisty fivesome clamber onto a bed and eavesdrop on a Cabinet meeting hosted by Hugh Grant. In the intentionally naff disaster movie *Sharknado 5* the palace moonlights for Buckingham Palace, in a scene in which the hero (called Fin, naturally) crashes through a window on the back of an airborne shark and down the grand staircase, before landing at the feet of the Queen (played, for reasons unknown, by the Spanish actor Charo), at which point her missing crown is retrieved gooily from the shark's jaws. You'd hope that Addington Palace charged plenty for that shoot; the movie apparently made $789 million in its first year.

The modern stereotype of Croydon points to a philistine, materialistic wasteland – a godless place. Yet here at Addington you see how long the town sat at the heart of the Church of England. Portraits of the six archbishops who made use of Addington Palace – Charles Manners-Sutton, William Howley, John Bird Sumner, Charles Thomas Longley, Archibald Campbell Tait and Edward White Benson – hang on the walls. Their remains (or those of five of them, anyway) are buried in the nearby church of St Mary the Blessed Virgin.

The last of these prelates, Edward White Benson, created the format of the Nine Lessons and Carols, bringing it with him to Addington in 1890. From here the tradition grew into the TV event broadcast to millions of viewers every Christmas Eve from King's College Chapel, Cambridge. But even in the modern era Croydon has, characteristically, led the way into the

THE WHITGIFT THAT KEEPS ON GIVING

Addington Palace: former home of the Royal School of Church Music, the place that inspired Henry James to write *The Turn of the Screw* and location for disaster flick *Sharknado 5*. © *Will Noble*

future. As early as 1985 Wilfred Wood became Britain's first ever Black bishop as the Church of England's Bishop of Croydon. It wasn't until 20 years later – two years after the end of Bishop Wood's lengthy tenure – that John Sentamu became Archbishop of York.

Edward Benson's ghost is said to have spooked the choristers at Addington Palace while it was home to the Royal School of Church Music (RSCM).

CROYDONOPOLIS

Five Archbishops of Canterbury are buried at St Mary the Blessed Virgin in Addington, although Croydon Minster has six. © *Will Noble*

This was based at Addington between 1954 and 1996, producing scores of eminent choristers and organists, while below stairs in the underground passages the publishing department churned out reams upon reams of *Te Deums*, *Magnificats* and suchlike, to soundtrack thousands of churches up and down the land. The RSCM moved on to Dorking in 1996, and later still Salisbury, and these days you're more likely to hear '*Chan Ve Shaukan Mele Di*' blasted across Addington Palace's Capability Brown-landscaped lawns. They do a roaring trade in Hindu, Sikh and Muslim wedding ceremonies – almost 2 per cent of Britain's Indian population lives in Croydon.

While still in corporeal form, Benson was co-founder of the Cambridge

Ghost Society, and relished telling ghoulish stories to palace guests. One wintry evening at Addington Palace, Benson entertained the American author Henry James with a fireside yarn about a group of depraved young children left in a big house who are haunted by their deceased servants. 'The vaguest essence only was there – some dead servants and some children,' the author later recalled to Benson's son.

> This essence struck me and I made a note of it (of a most scrappy kind) on going home. There the note remained till this autumn, when, struck with it afresh, I wrought it into a fantastic fiction which, first intended to be of the briefest, finally became a thing of some length and is now being 'serialised' in an American periodical.

James had adapted Benson's story into *The Turn of the Screw*.
Revenants are rampant in Croydon, if you happen to believe in them. Back at the almshouses, the spectre of John Whitgift's guardsman is said to float around after dark. The guardsman dozed off while keeping watch outside Whitgift's chamber, tumbled down the stairs and cracked his neck. Stairs, in fact, have a lot to answer for. In Croydon Palace a 'Green Lady' stalks the corridors, wailing terribly. Her ghoulish fate, so the story goes, came about after throwing herself and her baby down a staircase. A former pupil of the Old Palace School explained to *Inside Croydon* how to conjure up the Green Lady on demand: 'If you stamp on one of the stones five times . . . I think it was five . . . her ghost would appear and fall down the stairs again.'

Addington's church may have five archbishops buried beneath it, but Croydon Minster said 'Hold my communion wine' and bagged six. A parish church has stood in Croydon's Old Town since Saxon days, and by the late fourteenth/early fifteenth century was a spectacular Perpendicular Gothic structure in the style of Winchester Cathedral or Eton College Chapel.

Whitgift lies here among his predecessors and successors. The tomb of Gilbert Sheldon (Archbishop of Canterbury from 1663 to 1677) features his carved effigy laid out on his side, elbow on a pillow and head resting against the palm of his hand – not unlike that scene in *Titanic* in which Kate Winslet asks to be painted like one of Jack's French girls. Sheldon is probably best known today as the namesake of Christopher Wren's Sheldonian Theatre in Oxford. Not so well-known is the rumour – spread by none other than Samuel Pepys – that Sheldon 'do keep a wench, and that he is a very wencher as can be'. Pepys's unimpeachable source? His 'Cozen Roger'.

Sheldon's tomb is largely original, but it's one of the few things in the Minster that is. On the frigid night of 5 January 1867, the bellringers were having their annual dinner across the road in the Rose & Crown. Halfway through, William Kilmaster, the sexton and a ringer himself, nipped back over to the medieval church to attend to the stoves. On returning to the pub, he winked at the bellringers: 'It's a cold night, boys, but I've warmed them up all right.' He certainly had. Moments later, all that was left of Croydon's parish church was a smoking church-shaped shell. It was hastily rebuilt under the supervision of George Gilbert Scott, architect of the glorious Midland Grand Hotel that fronts St Pancras Station and the Albert Memorial in Kensington Gardens, and finally became a minster in 2011, part of one of Croydon's many grabs at city status.

Edmund Grindal had been the first archbishop to be buried at the church when he died in 1583, although after the fire no one bothered to replace his tomb, and he's now remembered with an undistinguished plaque. 'Clearly for Croydonians, Whitgift was a far more appealing figure than Grindal, and presumably that was still the case in the nineteenth century,' Andrew Bishop, the current Vicar of Croydon, tells me.

Whitgift's tomb, on the other hand, was painstakingly recrafted to the authentic Elizabethan plans. He lies in saintly repose beneath a stained-glass window celebrating his life's work, hands clasped in prayer, sporting a cherry-red frock. Two naked putti watch over him, one with a golden shovel, the other kneeling on a skull. Every Founder's Day in March, the residents of the almshouses, as well as Whitgift school students, descend on the Minster and place chaplets at Whitgift's feet.

It was once a tradition for Whitgift boys to pin dainty Star of Bethlehem flowers to their blazers as a visual nod to Elizabeth's 'white gift' pun, and to this day they wear white carnations. As an epitaph for Whitgift by Thomas Churchyard begins,

> Croydon can shew, his works, life, laud and all,
> Croydon hath lost, the Saint of that sweet shrine.

Chaplets are still laid at the tomb of John Whitgift in Croydon Minster, well over 400 years after his death. © *Will Noble*

He might be long dead, but you can't gloss over Whitgift's legacy, continued now by the Whitgift Foundation. It includes a portfolio of schools: Whitgift School in South Croydon, where boarders pay north of £50,000 a year; private boys' day school Trinity School of John Whitgift in Shirley; and the Old Palace School for girls near the Minster, although that will close in 2025. There are Whitgift Foundation care homes, too. In a roundabout way, then, Whitgift is watching over Croydonians from youth to old age. 'That's quite a legacy when you're over four hundred and twenty-five years on,' says Andrew Bishop.

In recent years Whitgift schools have provided us with some head-turning students. Derren Brown described Whitgift School as a world of 'peacocks and quadrangles', where he struggled to fit in. From this sense of awkwardness sprouted a need to perform, and Brown went on to conquer television, and later returned to Croydon (the Whitgift Centre, indeed) for a TV experiment in which he got dozens of shoppers to stick their arms in the air at the same time without realising what they were doing.

Even more fantastically, Whitgift School's library enchanted a young Neil Gaiman. While he was studying here in the 1970s, Gaiman spent as much time as he could devouring the likes of *Gormenghast* from the library shelves. It was also here, so the storyteller Patrick Ryan informs me, that Gaiman discovered Lucy Clifford's *The Pear Drum*, a sinister tale in which two wayward girls are landed with a 'new mother' with glass eyes and a wooden tail. Its inspiration is said to have stayed with Gaiman, who would go on to write his novella – subsequently to be made into a movie – *Coraline*.

John Whitgift, then, is the Whitgift that keeps on giving. Much of the land in the centre of Croydon that he purchased all those centuries ago is still owned by the Whitgift Foundation, including that on which the jilted Whitgift Centre itself stands. Altogether, the Foundation owns tens of millions of pounds' worth of Croydon real estate, which over the years has had a profound effect on Croydon's trajectory, and continues to do so. As Professor

Katrina Navickas, a historian of protest and public space, says, 'The Whitgift Foundation are always in the back of things.'

Whitgift and many subsequent archbishops made an indelible contribution to Croydon, never mind the country as a whole: according to *The Archbishops' Town*, nothing less than a formal urban identity. The modern town would be a very different place without them. Yet when Addington Palace was sold off in 1897, the archbishops' relationship with Croydon grew nebulous. Christianity was beginning to make room for a new object of worship: industry. Croydon was eager to snap free from the ecclesiastical apron strings, and make a go of things itself, using the kind of civic clout and scientific thinking that couldn't be achieved when you had a man of the cloth breathing down your neck.

And boy, did Croydon have some ambitious schemes up its sleeve.

4

THE IDEAL SITUATION

'We landed.'
'Where?'
'South Croydon. Hill View Road to be exact.'

If you'd been on a prim-looking housing estate in South Croydon in the summer of 1976, you might have seen a midnight-blue police box materialise out of thin air, and a young woman sporting candy-striped, star-spangled dungarees and a shaggy cream cardigan step out.

OK, you wouldn't. Not just because *Doctor Who* is widely considered to be a work of science fiction, but also because in this episode, 'The Hand of Fear', Tom Baker's erratic Time Lord has punched some iffy coordinates into the TARDIS and dropped off his assistant Sarah Jane at the wrong location. 'This isn't Hill View Road . . .' mutters the freshly-jettisoned sidekick as she surveys her surroundings and the time machine evaporates behind her. 'I bet it isn't even South Croydon . . .'

She's right too; some 30 years later the show finally delivered the pay-off, revealing that the Doctor had accidentally left Sarah Jane not at her aunt's place in Croydon but on the fringes of Aberdeen.

Still, at least Croydon was the *intended* destination, the end point, *home*. In real life, it is usually somewhere you pass through on the way to somewhere else – just as those archbishops were doing for centuries.

To heavily paraphrase Judy Garland, there's no place like Croydon. It inhabits a uniquely sweet spot, perched on the southern cusp of the greatest city in the world, while serving as a gateway to the rolling North Downs, the effulgent coastal south and, for those yearning for more exotic climes, Gatwick Airport, just 15 miles roughly south. 'The town's proximity to London places it in a most fortunate position,' brags a promotional pamphlet for Croydon's Grants department store in 1946:

> Frequent electric trains link it to main London stations, a journey which takes less than a quarter of an hour. In the other direction an equally frequent service of electric and steam trains transport travellers through the glorious countryside of Surrey and Sussex to the coast, to Brighton, Hove, Worthing and Eastbourne, in less than an hour.

Today any estate agent will furnish you with similar sales patter. 'You can get into Croydon very quickly,' Ashley Whitehouse, a sales manager at the local Foxtons, tells me, 'and get out of Croydon very quickly, both by train and car with complete ease.'

It's true. Croydon has an embarrassment of railway stations – East Croydon, West Croydon, South Croydon – plus a liberal scattering of others across the borough, all offering to take you elsewhere. There are mainline trains, Overground trains, trams, buses and Superloop services. Then there's the A23, a perma-pumping artery flinging high-speed traffic back and forth between London and the South. The only time this changes is during the early hours of a November morning once a year, when hundreds of pre-1905 vehicles participating in the Veteran Car Run pootle through Croydon at yesteryear speeds. Even then, none stop in Croydon, unless they happen to break down.

Croydon's throbbing through traffic has often proved an unwelcome distraction to the great and the good. At his home in Norbury, a residential neighbourhood just north of Croydon town centre, the composer Samuel Coleridge-Taylor was constantly discomposed by the trams, motor omnibuses and traction engines on the London-to-Brighton road, which 'rattled and rumbled by with brief intermission, all through the twenty-four hours', badly affecting his nerves, and giving him a recurring dream in which a grinning horse pulled a loaded vehicle up a slope. Still, at least he got *some* sleep. On the eve of her solo flight from Croydon to Darwin in Australia, the young aviatrix Amy Johnson was kept awake by traffic on the Purley Way (one of the country's first bypasses). It was not the ideal start to a voyage that would end up lasting nineteen frazzling days.[*]

Meanwhile, in the early 1900s an unfulfilled twenty-something teacher at Croydon's Davidson Road School called David Herbert Lawrence was preoccupied by Croydon's traffic in a different kind of way. During lunch breaks he would climb to the top floor of the school to watch the trains to-ing and fro-ing between Croydon and Norwood Junction. D. H. Lawrence, as he became better known, went on to use trains frequently in his poems and novels, observes the author Helen Baron, 'to coerce – overtly or subliminally – the reader's feelings and responses'. In 'Kisses in the Train', the incessant speeding of a train collides with Lawrence's trademark lustiness:

And the world all whirling
round in joy
like the dance of a dervish

---

[*] On 11 September 1877, anyone would have been distracted by the sight of two clowns sat backwards on donkeys, as they rode from London to Croydon.

did destroy

my sense – and reason

spun like a toy.

Everyone wants to *use* Croydon. They just don't necessarily want to stick around. 'It cannot be called a centre,' wrote Eric Parker in his 1908 book *Highways & Byways of Surrey*: 'for one returns to centres, and Croydon has little that would recall a traveller.' Though it's been described as a 'dormitory town' (that strange demarcation which suggests somehow it's not a *real* town), for much of its earlier existence Croydon was more of a corridor or pitstop. Ever since Roman times it had been used in this way; somewhere to refresh the horses/legionaries and get some kip for a few hours. The archbishops merely continued the tradition of using Croydon en route to somewhere else, as did pilgrims joining the famous Winchester–Canterbury route. But this 'rest stop' reputation was really confirmed in the mid-eighteenth century.

In 1750, Richard Russell published *A Dissertation on the Use of Seawater in the Diseases of the Glands, Particularly, the Scurvy, Jaundice, King's Evil, Leprosy and the Glandular Consumption*. It may not have had the pithiest title, but the treatise, which advocated bathing in sea water, as well as glugging a pint of it every now and again, became incredibly successful. So successful, in fact, that it turned around the fortunes of the clapped-out fishing town of Brighthelmstone, not yet known as Brighton. A whole faculty of dubious doctors was soon cranking out advice on thalassic health, and wellness-obsessed folk came tumbling into the Sussex resort to soak their bones in its briny waters.

Brighthelmstone's stock rose further when, in the 1780s, King George III's eldest son George (the one who'd later become the Prince Regent and ultimately George IV) began taking extended holidays here. Ostensibly this was to buff up George's own health by partaking in saltwater therapy, although if, as the TV historian Lucy Worsley claims, George was scoffing 'two pigeons and three beefsteaks, three parts of a bottle of Moselle, a glass of dry Champagne, two glasses of Port and a glass of Brandy' for *breakfast*, there's only so much good a paddle in the sea would've done him. Nevertheless, Brighthelmstone metamorphosed into Brighton, itself rebranded as the uber-trendy 'London-on-Sea', and Croydon, by dint of where it was on the map, became a glorified service station for the booming stagecoach industry.

In *Memorials of Croydon Within the Crosses*, John Ollis Pelton conjures up a Croydon High Street 'with here a slow, grey-tilted carrier's cart, and there a Brighton stagecoach, stopping to change horses, with the scarlet-coated guard on the back seat, equipped with post-horn and blunderbuss'. Guards on these stagecoaches tended to pack a piece due to the perils of highwaymen. This was the era of Dick Turpin and his gang – in fact, legend has it that Turpin lived in a cottage in Thornton Heath, a jot north of central Croydon, which had a secret staircase leading to the roof for swift escapes.

Whether or not this was true, the threat of highwaymen was very real. A roving brigand by the name of O'Brien literally defrocked a Croydon vicar, after winning his canonical gowns in a game of cards.[*] In 1795, Jerry Abershawe, aka the 'Laughing Highwayman', was tried and sentenced to death in Croydon, becoming the last hanged highwayman to have his body put on public display – and a popular attraction it proved, too. Just before he swung, Abershawe kicked off his boots, thus disproving his mother's prediction that he'd die with them on.

The threat of footpads patently didn't deter most holidaymakers, because by 1815 some 100,000 were shuttling along the Brighton Road, and Croydon was ready for them. Inns like the Greyhound, the King's Arms,

---

[*] The vicar, one William Clewer, was a notorious figure in Croydon, an Edmund Grindal on steroids. Clewer plagiarised other vicars' sermons, stole from bookshops, extorted money from parishioners, regularly got steaming drunk and once kidnapped a fourteen-year-old, and so thoroughly deserved his defrocking.

the Crown and the George provided longer-haul travellers with somewhere to sleep before they got a wriggle on the next morning. The Greyhound had a large gallows sign straddling the entire street, much as a motorway sign might indicate a service station today. At the George, a nasty legend persisted that lodgers were murdered in their beds, with their remains boiled in a cauldron by the landlady, aka 'Old Mother Hotpot'. (Another, more believable, story tells of an 'Old Mother Hot*water*', a whizz at cooking and cleaning, who didn't do away with any of her guests.) The Prince Regent himself once called in at Croydon on his way to one of his Brighton jollies, only to be booed and hissed at by a mob outside the Crown, of all places. Taking the altercation to heart, George made a point of never coming to Croydon again.

Early-nineteenth-century roads were sticky with mud and rutted with potholes. Even though the stagecoaches, given dynamic names like *Comet*, *Vivid* and *Dart*, only went about 10 miles an hour at top whack, it still made for a skull-rattling ride. Turnpike trusts were set up to improve the routes, and toll gates became commonplace, raising much-needed money for the roads' upkeep. One of these toll gates was almost the death of a sitting prime minister. Riding back from a party hosted by Lord Liverpool at the Addiscombe Place mansion one evening, a squiffy William Pitt the Younger (likely to have sunk three bottles of port, as was his daily intake) charged straight through a gate at Croydon Common and was shot at by the nightwatchman, very nearly becoming the only British prime minister to be assassinated by accident. The toll-dodging PM, as it happens, is the same man responsible for introducing income tax to Britain.

Some industrious Croydonians pivoted to manufacturing carriages. Businesses like William Waters and Messrs Lenny and Co. churned out broughams, phaetons, clarences, wagonettes, stanhopes and Parisians. Eighteenth-century Croydon was the Motor City of its time – just minus the motors. The town even made its own brand of carriage, the Croydon

Basket, which was, bragged Messrs Lenny and Co., 'so famous throughout the world' that they'd had to 'greatly enlarge their premises'. In addition to their showroom in Croydon, the company had another in Park Lane, Central London – the same spot where BMWs and Aston Martins now beckon to petrolheads through the plate-glass windows. Croydon had dipped its toes into a love affair with wheels it's still mixed up in.

It wasn't long, though, before roads were falling out of fashion, the efflorescence of railway mania in the mid-1830s bringing Croydon's carriage trade to a jarring halt. Except here, Croydon thrusts a spanner into the spokes of history. As early as 1803 it already had a railway.

Croydon had one of the first ever public railways.[*] Admittedly the Surrey Iron Railway didn't have any steam engines – or, come to think of it, passengers. But it was the first company in the world to use the word 'railway' in its title. With the Industrial Revolution already going full tilt at the dawn of the nineteenth century, the industrialists orbiting Croydon, and all along the Wandle Valley from Croydon to Wandsworth, knew they needed to link up with London and the Thames to grab a slice of the steaming pie. Though technically Croydon was joined to the Thames by the Wandle, a river that since the 1600s had powered a thriving textile mill industry, this wasn't navigable.

The decision was taken, therefore, to build the Surrey Iron Railway, a line running between Wandsworth in South London and Croydon (plus a fork that went south-west to Hackbridge, near Mitcham Junction). The railway was then extended from Croydon southwards down to Godstone in 1805, as the Croydon, Mertstham and Godstone Iron Railway.

As railways went, the Surrey Iron Railway and its extension were primitive. Open wooden wagons were set onto L-shaped iron rails (technically

---

[*] The 'public' aspect of the railway was that anyone could pay a fee to put their own wagons and horses on the rails. Tolls depended on what you were carrying, and how much: dung was 1d per ton per mile, whereas other, fancier, types of manure cost 2d.

'plates', as any railway pedant will tell you), and pulled by horses. Yet it was revolutionary. The rails meant horses could pull substantially heavier loads, and for longer distances than they'd been able to do on any road. This theory was put to the test after the railway's extension was completed, and a group of men gathered at the Fox pub in Hooley to see out a wager by one Mr Banks. The railway's engineer William Jessop had claimed a horse could pull five tonnes in one direction and 10 in the other, but Banks reckoned a horse could actually pull a 36-tonne load along the track for six miles.

The horse was chained up to 12 wagons, each packed with three tonnes of stones, and duly dragged the load for six miles with relative ease. Its reward wasn't a lump of sugar but being made to carry *more* wagons back the other way, this time loaded up with *50 workmen*. Banks won his wager with a flourish, and the workmen unwittingly became some of the first passengers ever to travel by rail – some two decades before the Stockton and Darlington line (touted as the world's first passenger-carrying railway) opened.

The Surrey Iron Railway was built in the first instance because William Jessop had deemed a canal between Croydon and London unworkable due to a lack of available water. The mill owners dotted along the Wandle were particularly reluctant to relinquish any of their valuable power supply. Even so, the industrious folk of Croydon soon decided they were going to build a canal anyway. After all, now that Croydon had a railway, materials could easily be shuttled to the canal for an onward journey to the London Docks.

And so, on 22 October 1809, thousands of Croydonians swarmed around the canal basin at West Croydon to greet a motley procession of barges festooned with flags, which had sailed down from Sydenham along the freshly dug canal. With the patriotic gusto of the time, gun salutes were fired, 'God Save the King' was sung, and church bells rang out. A Mr J. (T.) Welsh performed a specially penned ditty:

THE IDEAL SITUATION

All hail this Grand (Great) day when with gay colours flying,
The barges are seen on the current to glide,
When with fond emulation all parties are vying
To make our Canal of Old England the pride.

This was a huge moment for Croydon. Overnight, it became the first town in the country with both a railway *and* a canal. What's more, it was now umbilically linked to London.

The waterway, which ran from Croydon up to the Surrey Docks at Rotherhithe, would shuttle the fruits of the Surrey countryside and forests – timber, clay lime, flour, flint, seeds – to the city, and in exchange, Croydon would receive shipments of that inexhaustible fuel of progress, coal. (By this time, charcoal was no longer de rigueur.) The masterplan didn't end there. As the canal's investors decamped to the Greyhound for wine and back-patting, escorted by an army of workmen marching with their tools on their shoulders, they were already hatching a vision to extend the canal down to Portsmouth. War was raging against the French, and the Croydon Canal would offer a way of shuttling naval supplies and ship timber to Portsmouth, without exposing them to rough weather and cannon-happy foes.

Sadly, the party was short-lived. Canal NIMBYs came crawling out of the woodwork: a group of landowners along the proposed Portsmouth extension decried what they opined would be a 'useless ditch'. One Robert Marshall produced a pamphlet claiming that as the first part of the canal had taken so long to dig, it'd be 20 years before the Portsmouth leg was completed. Mill owners complained about the only thing they knew how to complain about: that an extension would steal *even more of their precious Wandle water.*

Things weren't looking too rosy for the stretch of the Croydon Canal that already existed either. The canal took an age for boats to navigate: from start to finish, there were 28 locks to tackle, and because of this traffic jams were commonplace. Banks on either side of the canal began collapsing too.

CROYDONOPOLIS

A lack of water supply meant boats, already sitting low in the canal, could often only carry half loads.

What's more, the pamphlet-pushing Robert Marshall had been right: the canal *had* taken far longer than it should have to build, not to mention costing three times more than budgeted. There were never sufficient funds to see it reach its full potential, and investors never saw a return on their money. The whole project was a catastrophe, to the extent that an extended essay on it is entitled baldly, *The Canal That Failed*.

For all its litany of flaws, though, the coup de grâce for the Croydon Canal (and the Surrey Iron Railway, which ceased operating in 1846) was the coming of the real, full-fat railway revolution – this time minus the horses, and with fare-paying passengers to boot. As the doomed waterway became one of the first in Britain to be ousted by trains (it closed in 1836), a local by the name of F. L. Selous conjured up a curious poem that imagined a conversation between the canal and the railway. At one point the railway crudely sneers to the canal,

> Thou dozy old god – the water 'tis plain,
> (What little is left) has got to your brain.

Just 27 years after the pomp and ceremony of its ribbon cutting, the majority of the Croydon Canal was sold off to the railway companies for £40,259. As Selous concluded in lilting melancholy:

> The magic steam whistle has sounded his knell,
> And the spirit is lost of the Croydon Canal.

Most of the canal was soon infilled with rail lines – recycled for the next generation of transport.[*] Here was Croydon setting the precedent for what

---

[*] These are the same lines in operation today; when you catch an Overground train from West Croydon to New Cross Gate, you're tracing much of the route of the old canal.

would become a tendency: to pounce on something pioneering, like the canal, then ditch it just as quickly for something better. But certain bends in the canal were impractical to adapt. Croydonians jumped at the chance to use these leftover sections, and fishing, swimming and boating became popular pursuits. 'Most picturesque it was,' wrote William Page in his memoirs of Croydon. 'You might almost fancy yourself on a small river in the interior of Africa being so wild and uncultivated.' When the water froze over in winter, fun seekers skated all the way from Croydon to Brockley, where the locks stopped them getting any further.

Though they'd hastened the demise of the Croydon Canal, the railway companies actively encouraged passengers to use what was left of it for riparian pleasure. 'Marquees are erected in the Wood close to the Anerley Station,' ran an announcement in 1840's *Robinsons Railway Directory*, 'and Parties using the Railway will be permitted to angle in the adjacent canal which abounds in fish.' Elsewhere, waterside pubs flourished on the former canal. A Swiss hotel was built on the banks at Anerley, serving teas and encouraging visitors to negotiate its maze. In South Norwood, the Jolly Sailor pub was celebrated for its tea gardens running down to the water. It would later give its name to the local railway station before it became Norwood Junction.

Another pub that took advantage of the repurposed pieces of the canal was the Anerley Hotel, which opened in 1841. It still exists as the Anerley Arms, a Samuel Smiths pub festooned with Sherlock Holmes paraphernalia. This isn't just some whim of the proprietor: Arthur Conan Doyle used to live in South Norwood (often cycling around with his wife Louise on a tandem tricycle), and some of his Sherlock stories, including *The Sign of the Four* and 'The Cardboard Box', are set in this part of Croydon. It was also while in Norwood that Conan Doyle killed off Holmes in the infamous Reichenbach Falls story, before realising there was life in the old dog yet and figuring out a way to resurrect the deerstalker-sporting detective. The Anerley Arms itself features in 'The Adventure of the Norwood Builder', in which a young lawyer is accused of murdering a builder, and Holmes sets about proving otherwise, eventually inveigling a confession when he raises a false alarm:

> 'Fire!' we all yelled.
> 'Thank you. I will trouble you once again.'
> 'Fire!'
> 'Just once more, gentlemen, and all together.'
> 'Fire!' The shout must have rung over Norwood. It had hardly died away when an amazing thing happened. A door suddenly flew open out of what appeared to be solid wall at the end of the corridor, and a little, wizened man darted out of it, like a rabbit out of its burrow.
> 'Capital!' said Holmes, calmly.

Real-life murder mysteries weren't uncommon on the Croydon Canal. Its secluded stretches were a honeypot for drownings and suicides. The most perplexing case was the discovery of the body of a woman in her mid-twenties in the canal on 29 May 1831. She was dressed in a black silk gown and straw hat, had blackened eyes and a fractured forehead. She was also pregnant. It was discovered she had been into a shop on Sydenham Common with a female friend the previous day, announcing to the shopkeeper that she was going to meet with the father of her expected child that afternoon and tell him the news. The two women were later spotted out on the canal with two men, in two separate boats. The next morning, the boats were found abandoned. Murder was suspected, but then a local policeman muddied the waters by revealing he'd encountered the deceased woman by the canal three weeks previously, where she'd threatened to commit suicide. It doesn't seem the men were ever located, and the conclusion of this strange episode is lost in the murk of time. Where was Sherlock Holmes when you needed him?

You can still find a few glistening fragments of the Croydon Canal here and there. Betts Park in Anerley contains a trough of water, around 175 metres long, where mallards pootle up and down and a few lucky residents overlook what is the only part left of the canal still remotely canal-like. Just west of here is South Norwood Lake, once a reservoir that fed the canal. Now, anglers flump back in camping chairs on little wooden jetties around the lake, occasionally reeling in a flapping tench or bream, while yachts from the Croydon Sailing Club swish around. It mirrors what Croydonians were doing for fun almost 200 years ago.

5
'NARNIA IN URBAN GREEN'

The ancient archiepiscopal town of Croydon lies at [your] feet; the Banstead Downs, in all their beauteous variety of fallow field and grassy meadow, are in the distance.

Croydon? Does this sound like the one place in the entire country Tom Chessyre would not come back to? It's time to tackle Bowie's 'concrete hell' slur head on.

Natural beauty is bountiful in Croydon. Yes, on Google Maps' satellite view the town centre looks like a coarse grey scab. But lapping at its fringes are blankets of green. A quarter of the borough – some 5,400 acres – lies within the Green Belt. Its boundaries are tickled by Surrey and the North Downs, where D. H. Lawrence spent his weekends gathering blackberries or primroses while pondering love, life and death. Croydon itself is studded with sweeping golf courses and great country piles, like those at Heathfield, Coombe Cliff and the old Selsdon Park Hotel. From the 1950s to the 1980s, Green Line coaches puttered through this verdant realm on prodigious routes traversing the capital from Kent to the Chilterns. The borough has over 120 parks and open spaces. Park Hill Gardens, with their candyfloss-like cherry blossoms. The chalk cliffs of Riddlesdown. The grassy sweep of Lloyd Park, where frisbees sail through the skies during summer games of disc golf. Croham Hurst Woods, laced with ancient beech and birch trees. The gorsy climes of the Addington Hills. To put into perspective just how arboreal the borough is, in the Great Storm of 1987 some *75,000* trees were uprooted.

Wildlife is abundant. The Croydon Beekeepers Association makes its own Croydon Honey, and even the RSPB has its roots in Croydon, thanks to the formation here of the Fin, Fur and Feather Folk in 1889 by Croydonian Eliza Philips, and the wonderfully named Etta Lemon, who became known as 'the Mother of Birds'.

Ameena Rojee is an artistic photographer who has captured Croydon's comeliness on camera, and explains to me how it can surprise and delight, with a story about visiting Roundshaw Downs, to the south of the town.

> I went in by foot, alone, in the thickest fog I've ever seen ... Only the dark streaks of crows in the greyest of grey fogs. Then I saw a black shape on the horizon. I get closer, and I see that the shape is a bull. Slowly, I see more of the most gorgeously soft brown cattle appear in the dense fog, and I'm wondering to myself what the hell are a bunch of bulls doing in Croydon? It was a moment of pure awe and joy, and for me completely encompasses the duality of Croydon.

It's no wonder that in Rojee's book, *Crocus Valley*, Croydon's Poet Laureate Shaniqua Benjamin describes the area as a 'real-life Narnia in urban green, fulfilling atmospheric stories of every day and its everyday people; this is a world beyond structured things.'

Croydon might be grey, but it is also very green. And purple. 'Crocus

Park Hill Park in East Croydon, one of more than 120 parks and open spaces in the borough. © *Will Noble*

Valley' is where Croydon gets its name from.* Wild purple crocuses once flourished in this basin at the foot of the North Downs and along the Wandle Valley. When Croydon was a Roman staging post between Novus Portus (Portslade) and Londinium, these crocuses were harvested for their saffron, which was infused in medicines, spiced honey wine and hot baths. By the time the Anglo-Saxons were around, they'd named the place *Crogedene*. It's an unlovely-sounding name which, when translated, blossoms elegantly into 'Valley of the Crocuses'.

---

* That's one of the theories, anyway: others suggest it derives from 'crooked valley' or 'chalk hill'.

CROYDONOPOLIS

The name 'Croydon' romantically translates as the 'Valley of the Crocuses', a flower which continues to blossom across the borough every spring. © *Will Noble*

That might sound a bit rich for a town peppered with manmade megaliths, but even now the borough sparks up with lilac-coloured crocuses each spring, from the Minster churchyard to the front lawns of Coulsdon's Tudorbethan villas.

Croydon's history is heavily scented with petals: Cicely Mary Barker penned her Flower Fairies books here, a series of poems and illustrations about Tinker Bell-esque sprites who live inside plants. Maybe Barker's Crocus Fairies with the 'Crocus of yellow, new and gay; Mauve and purple, in brave array' took their cue from Croydon's relationship with this genus of flower. Certainly many of the artist's young models were Croydonians: she drew children from her sister's nursery, or simply those she saw playing in the grubby Croydon back streets. There's a memorial garden to Cicely Mary Barker secreted in Park Hill Gardens near East Croydon Station – only flowers and trees that can be found in her books are planted here.

Lavender also flourished in the sandy soil around Croydon. In 1883 the French agriculturalist and chemist Phillipe Augustus Lelasseur built a steam-powered distillery on Mitcham Road in West Croydon and began alchemising essential lavender oils under the English pseudonym 'John

Artist Cicely Mary Barker lived in the Waldrons near Duppas Hill and used Croydon children as models for her famous Flower Fairies books.

Jakson'.[*] Mitcham Lavender Water was an instant hit, and at the Exposition Universelle of 1885 in Paris the Croydon-distilled fragrance beat off the French competition to win a gold medal. 'It is astringent, stimulating, improves the complexion and renders the flesh firmer,' ran one of the ads.

One green space that no longer exists, however, is the Beulah Spa, a nineteenth-century Xanadu built on the precipice of Croydon, to sublime effect.

Taking the waters wasn't just a Brighton fad. In Bath, wellness-conscious popinjays and their squeezes were knocking back glasses of sulphurous water. London itself was bestrewn with pleasure grounds, from St Chad's Well in Camden, where punters were charged 4d for an elixir that would 'cure the Scurvy, Bile, Worms, Piles, Indigestion, Nervous Complaints, Seminal Weaknesses, and various other Disorders too numerous for an advertisement', to Streatham with its delightful Rookery Gardens, which exist to this day. To the south-east of Streatham, perched at the tip of modern-day Croydon, stands Beulah Hill, where in 1831 the landowner John Davidson Smith let loose a lame horse, with astonishing consequences.

The liverish creature, so the tale has it, having slurped from the Beulah Hill spring, sprang back to life, growing fat and glossy: there was something life-giving – *miraculous*, even – in that water. Again, Croydon's geographical location was nudging the course of its fate.

Though the story about the horse is likely to be semi-apocryphal, it's true that Smith sent off samples to the scientist Michael Faraday, who immediately declared the libation more efficacious even than the spring water at Bath or Wells. Rich in magnesium sulphate, here was a panacea on tap. Smith wasn't going to look a gift horse in the mouth, so to speak, and wasted no time commissioning Decimus Burton, a bright young architect who'd later

---

[*] Some say the savvy entrepreneur purposefully spelled it this way so people remembered it.

'NARNIA IN URBAN GREEN'

design the Palm House at Kew and the layout of London Zoo, to create Croydon's take on the Garden of Eden.

Chris Shields, author of *The Beulah Spa 1831–56*, was born on Beauchamp Road, which overlooks the site of the old spa, and has long had a fascination with this spot. 'The fact that this suburban area was once a magical pleasure garden aimed at high society still amazes me,' he tells me.

> I think the most exciting thing for me about it all was that the visitor, when travelling from the cramped and dirty conditions of South London, would suddenly be hit by a fantastic view of the countryside, going on and on forever, and you'd see green grass, flowers and rustic buildings – grottos, arbours, benches, tents and marquees . . .

You entered via the Hansel-and-Gretel-esque Tivoli Lodge, to be handed a bottle of the chalybeate water and let loose into a utopian array of gardens, woodlands and lakes. American aloe, geraniums and fuchsias exploded in a riot of colour and perfume. A brass band played on the lawns daily. Johann Strauss Senior once conducted here. There was archery (bow and arrows provided), you could flick through the latest Mary Shelley tale in the octagonal reading room, or seek out the handsome Beulah Minstrel, who drifted about in a cloak and turban serenading guests with guitar ballads and returning measly tips with a graceful bow.

Paramours chased each other around a maze before dancing waltzes and polkas in the rosary. Pablo Fanque – of the Beatles' 'Being for the Benefit of Mr Kite' fame – performed a tightrope act. On any given day at the Beulah Spa you might see jugglers, acrobats, opera singers, sword swallowers, even newfangled balloonists sailing into the heavens. As dusk set in, fireworks detonated in the sky. It was wild. It was sensational. It was paradise. It was Croydon.

Another of the pleasure park's natty attractions was a camera obscura.

From up here, thanks to the words of 1834's *A Guide to the Beulah Spa*, with which this chapter began, we can get a pin-sharp lie of the land:

> Further yet the scarcely perceptible towers of Windsor Castle give variety to the landscape; and the extreme distance is bounded by the dark blue outline of the Surrey and Hampshire Hills. Turning to the left you enjoy a view of Addiscombe Place, the seminary for Cadets of the Hon. East India Company;* of Shirley ... of the Addington Hills, clothed with heaths; and of the seat of His Grace the Archbishop of Canterbury, when the prospect deepening in extent stretches as far as Knockholt Beeches, near Sevenoaks, and winding round comprehends the tall spire of Beckenham Church, piercing through the dense woods which surround it, Shooter's Hill, Blackheath, and the Villages that intervene.

The well itself was covered in a thatch roof, not unlike a modern tiki bar; this was Decimus Burton's nod to the collier huts of the surrounding Great North Wood. As well as being the tenebrous workplace of many a Croydon collier, the woods were historically inhabited by Gypsies. Samuel Pepys recorded his wife visiting the woods in 1668 to have her fortune told, but by the time the Beulah Spa opened the trees had significantly thinned out, many Gypsies now plying their trade around Norwood. Among them was Floriana Cooper, known as the 'Norwood Gypsy Queen', who read palms in a tent at the spa.

As business at the spa soared, people flocked from across the country

---

* The Addiscombe Place mansion house and its sprawling grounds were home to Charles Jenkinson, the 1st Earl of Liverpool, and in the early nineteenth century purchased by the East India Company and turned into a military academy. From 1809 to 1861 it trained up young men – and indeed boys as young as 13 – to fight abroad in the company's private army. But cadets didn't need to travel far for a swedge. Thomas Frost recalls over a dozen of them getting into a scrap at a local fair with a travelling theatre company. 'The consequence', wrote Thomas Frost, 'was a sharp scrimmage, ending in the arrival of several constables, and the removal to the station-house of as many of the cadets as could not escape by flight.'

'NARNIA IN URBAN GREEN'

The Beulah Spa, designed by Decimus Burton and boasting a source of spring water deemed more efficacious than that of Bath or Wells ... in Croydon. *Alamy*

to visit this 'Versailles of London'. It drew in aristocrats, celebrities and royals. Victoria came to the spa four times indeed, as both princess and queen. During a *fête champetre* held for the Napoleonic war hero Marshal Soult in July 1838, a queue of carriages stretched a mile-and-a-quarter down the road.

There was the potential to build another Bath, right here in Croydon. Plans were drawn up to construct a sweep of palatial apartments – not unlike Bath's Royal Crescent – along the brow of the hill. It was dubbed 'the New Town of Beulah', and would have been a jewel in South London's architectural crown.

Instead, the Beulah bubble burst. Even when it opened, the fashion for pleasure gardens and spas had been waning. Uncharacteristically, Croydon had been behind the curve. As customers began drying up, various owners tried and failed to keep it profitable. But in 1854, the spa's fate was no longer in any doubt, as a new and extraordinary adversary glinted on the horizon.

Three years after it had hosted a third of the British population in Hyde Park, the great glasshouse that was the Crystal Palace was reassembled in bulked-up form in Sydenham, just across the border from Croydon in what would become the borough of Bromley. The Beulah Spa's crystalline new neighbour towered over it, its ludicrous fountains spurting over 75 metres into the air, the spray drifting over the spa like the blowing of an almighty raspberry. Queen Victoria and her subjects had a new toy to play with. She was clearly amused by the Crystal Palace too, visiting 23 times, putting her four Beulah Spa jaunts to shame.

Within two years of the Crystal Palace's arrival in South London, Beulah was abandoned and nature gradually reclaimed it. In 1949 Alan R. Warwick visited the site, finding a few crumbling vestiges and almost falling into the well. He described the water as 'dark, still and bitter-tasting'. Hidden in the long grass you'll find a granite memorial stone marking the site of the old well, placed here in 2018 by Chris Shields and the Beulah Spa History Society. From this bowl-shaped hill, the panorama at which so many fancy-free Croydonians gazed before remains beguiling, if very different. You won't spy Windsor Castle or Addiscombe Place now, but you can certainly see the IKEA chimneys: two bricky fingers thrust into the face of history.

With the well dried up, all that's left for you to do is go to the bar of the Beulah Spa Harvester, sip on a pint of sparkling water and muse on the fact that, in March 1966, this part of Croydon became fleetingly famous once again, when a dog called Pickles sniffed out a stolen object that had been stashed beneath a laurel bush. It was the World Cup.

The Crystal Palace may have put paid to the Beulah Spa, but it also gifted Croydon one of its most famous exports. In 1861, seven years after the

Crystal Palace was re-erected in Sydenham, the Crystal Palace Company's cricket team pivoted to play the trendy new game of football, and did so with the magnificent glasshouse as a backdrop, earning them the nickname 'the Glaziers'. A couple of years later, Crystal Palace FC were sat around the table at the Freemasons' Tavern in Holborn, as the inaugural set of rules for the Football Association were bashed out.* Palace swiftly built a reputation as a club to be reckoned with, helped by a 7–1 win over Chelsea in the 1906 FA Cup. (Sadly, Palace never appeared in any of the FA Cup Finals hosted at their home ground between 1895 and 1914.)

At the time, technically Palace weren't yet a Croydon club, but following the First World War they were turfed out of Crystal Palace Park by the Admiralty and, after borrowing West Norwood's ground for a spell, moved in 1918 to 'the Nest', the former ground of the disbanded Croydon Common FC.

Many fans who turned up to watch Palace at the Nest opted not to shell out for a ticket, instead catching a train to Selhurst Station and watching the game from Platform 1, a very decent vantage point. Palace left this ground in 1924, subletting the Nest to a team called Croydon Tramways FC. Today, thousands of football fans still alight from Selhurst Station en route to watch Palace play at Selhurst Park, their home for a century – and a ground built, believe it or not, for just £30,000.

By the 1950s and 1960s Palace were the pride of Croydon, winning promotion after promotion. In 1962 Real Madrid came to Selhurst Park, narrowly beating the home side in a 3–4 thriller. Ten years later Palace saw off Man United in a mud-caked 5–0 rout ('Will this be five? It's gonna be five! It *is* five!') still talked about now. The era heralded a striking rebrand for the club: in 1973 its larger-than-life, fedora-wearing manager Malcolm

---

* Of the 11 teams in that seminal meeting on 26 October 1863, Crystal Palace are the only still playing top-flight football. Others included Civil Service, Crusaders, Forest of Leytonstone and N.N. Club, which stood for 'No Names'.

CROYDONOPOLIS

Allison borrowed the nickname of Benfica, *As Águias* (the Eagles), and jooshed up the kit to blazon Palace's now-famous red and blue stripes. Though an eagle has dominated the club's crest ever since, the Crystal Palace itself has remained on it too, acting as the bird's glassy perch. Recent years have seen Palace restored to the top flight and managed – twice – by a Croydonian through and through: Roy Hodgson.

For all Crystal Palace's moments in the sun, including four league titles, it's another home game against Man United that sticks in the public consciousness, and for all the wrong reasons. It was in the forty-fifth minute of Palace's clash with United at Selhurst Park on 25 January 1995 that the French striker Eric Cantona was shown a red card for kicking the Palace defender Richard Shaw. 'There's the morning headline,' declared the TV commentator as Cantona stalked off, unaware that the *actual* headline was coming a few seconds later – at the very moment Cantona's boot studs connected with the face of a Palace fan. Jonathan Pearce was commentating on the radio, and his eruptive reaction (*'Oh, my goodness me! He has kicked ... He has punched a fan! Eric Cantona has jumped in and scissor-kung-fu-kicked a fan ....!'*) is chiselled into the annals of sporting (or should that be unsporting) history.

After Cantona's flying kick landed on Matthew Simmons, the two collapsed into a sloppy punch-up, and the world of football was disgraced. Cantona returned to the borough for sentencing at Croydon Magistrates Court where, to audible gasps, he was initially handed a two-week jail term. Incredibly, the kicking wasn't over – but this time it wasn't Cantona's doing. After being found guilty of two charges of using threatening words and behaviour, and banished from football stadiums for a year, Matthew Simmons decided the best course of action would be to kick the prosecuting counsel Jeffrey McCann in the chest. The splenetic Palace fan received seven days in prison, while Cantona's punishment was downgraded to 150 hours of community service.

The score of the match on that fateful day (as few will remember) was 1–1. Palace's equaliser came from a 24-year-old Gareth Southgate, who'd go on to suffer his own ignominy the following year thanks to a penalty miss against Germany in the Euro semi-finals. Though he hailed from Crawley, Gareth Southgate had studied at Croydon College, where he acquired a City & Guilds Certificate for Recreation and Leisure Industries. In 2018 Southgate would recapture English hearts as the manager of the England team whose quarter-final victory at the 2018 World Cup put Croydon back in the limelight with an orgasmic outburst of elbow-jerk lager-spraying by 2,000 fans. These soaking scenes at Croydon Boxpark were broadcast all over the world, and the venue – and its sticky celebration – have since become synonymous with the euphoric sensation of England scoring a goal.

6

WELL TRAINED

The same year that the Croydon Canal was filled in, Londoners climbed out onto their rooftops en masse to witness their city's first steam-powered railway line, the London and Greenwich Railway, whoosh into action. Conveying passengers over a viaduct comprising 60 million bricks and 878 arches, the great feat of engineering heralded endless new possibilities for travel. Two-and-a-half years later, at the Corbett's Lane junction in Bermondsey, the London and Croydon Railway tacked onto the London and Greenwich, linking Tooley Street (now London Bridge) to Croydon Station (now West Croydon).

Then in 1841 the London and Brighton Railway opened, parallelling the historic stagecoach route in linking Croydon to the voguish South Coast, and in 1846 this line merged with the London and Croydon Railway, becoming the London Brighton & South Coast Railway. Croydon was now connected to the capital, and the south, like never before – a transportational lynchpin, this time as part of the Southern Railway network.

But while steam engines seemed the obvious way forward, this wasn't a done deal – not as far as the London and Croydon Railway was concerned.

CROYDONOPOLIS

Croydon's experiments with the Surrey Iron Railway and the Croydon Canal had both been abject flops, but this time Croydon was *certain* it was onto the next big thing.

In 1844, three neo-gothic spires sprung up at West Croydon, Portland Road (today's Norwood Junction) and Dartmouth Road (Forest Hill). These weren't churches of religion, but science, and the experiment in question was atmospheric propulsion. The 'steeples' of these curious buildings (Forest Hill's climbed an impressive 120 feet into the air) were elaborately disguised chimneys for pumping stations. Each station created a strong vacuum which ran through a special track – one with an iron tube set in the centre of the rails – laid between Croydon and Forest Hill. The vacuum in the tube propelled a piston, which itself was attached to the bottom of a train. The idea, trailblazed by the engineer Samuel Clegg and the brothers Jacob and Joseph Samuda, was nothing short of ingenious: an *atmospheric railway*. The trains themselves created no smoke or dirt, and none of the huffing and chuffing of steam engines. Trains could run up steeper inclines, which meant building less track. What's more, with the atmospheric railway there was no chance of a collision, because only one train could run on a line at a time.

Initial trials of the atmospheric railway were promising, with trains said to hit speeds of up to 70 miles per hour (one account even suggests *a hundred* miles an hour). The futuristic trains went at such a pace passengers described the sensation as 'falling from a height'. 'If they can travel five miles they can travel 5,000,' exclaimed one newspaper. And the scientific breakthroughs didn't stop there. The atmospheric railway saw the construction of the first ever railway flyover – a 'flying leap' to cross the existing Brighton line just south of Jolly Sailor Station. A similar design is now used on railways globally. As with the canal, there was soon talk about extending the atmospheric line to Portsmouth.

The concept was enough to catch the attention of the eminent mutton-chopped engineer Isambard Kingdom Brunel, who went to see Croydon's atmospheric railway in action (along with another of the Samuda-Clegg

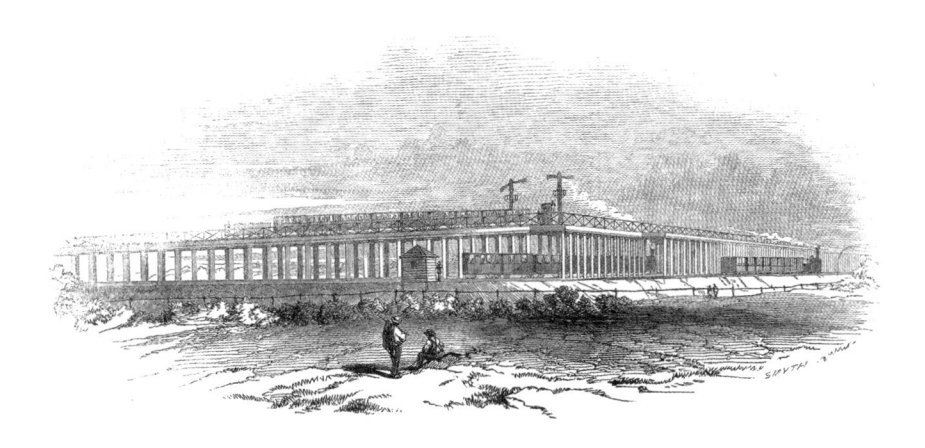

The Croydon Atmospheric Railway was a clever, quixotic but ultimately doomed rail project, cursed by everything from rats to rain. *Alamy*

lines operating near Dublin). Brunel was so impressed he decided to have a stab at atmospheric propulsion himself, on a coastal line between Exeter and Newton Abbot. While the world worked itself up into a sooty tiz over steam power, Croydon was pioneering a cleaner, smoother way forward – a Maglev for the Victorian age. 'The little Croydon frog,' writes Charles Hadfield in *Atmospheric Railways: A Victorian Venture in Silent Speed*, 'was indeed puffing itself very big.'

Unfortunately, the little Croydon frog was going to puff itself out altogether. The atmospheric railway had a snag list almost as long as the line itself. Air leaks meant that sometimes there wasn't enough propulsion to get the whole train over the flying leap at South Norwood, and that the back section of carriages had to be heaved over with the aid of a rope. When it rained, the brakes didn't function properly, trains slid past the stations and, because reversing wasn't an option, passengers had to clamber out along the tracks. This wasn't the only indignity they suffered. Male passengers were sometimes asked if they wouldn't mind rolling up their sleeves and *giving a bit of a push.* To add insult to injury, the train would sometimes then pick up speed so quickly it shot off without them.

Then there were the rats. They got a taste for the tallow and wax-slathered leather flaps maintaining the vacuum seal, and climbed inside the iron tube to lick and gnaw this off. Their reward, more often than not, was a shortcut to rodent heaven. The naturalist Frank Buckland reckoned hundreds of rats went flying through the pipe – sucked into oblivion – every day. Men in the pump house engine rooms even put sacks over the inlets to catch the incoming furry deluge.

Far more concerningly, *people* were getting killed. The *sotto voce* passage of the piston-driven trains led to some hairy encounters for passengers crossing the tracks (as with the electric cars of today), and in February 1847 an engineer named Thomas Smith was fatally struck. As if the railway gods were trying to tell Croydon something, a huge fire broke out at the Croydon end of the line, putting paid to the locomotive depot and carriage sheds, while the railway tracks themselves were 'twisted into fantastic shapes and divers forms'. Wrote one detractor of the atmospheric railway in 1845: 'We unhesitatingly assert, that hitherto it has proved, not only commercially, but practically, a decided failure, and Mr Samuda knows it.' (They must have been talking about the surviving Samuda brother. In the same year the atmospheric railway opened, an iron steamboat called *Gipsy Queen*, which Jacob Samuda was working on in the dock at Blackwall, exploded. He was scalded to death, along with six other men. You wonder what else he had percolating in that brilliant brain that never had the chance to see daylight.)

Samuda wasn't a Croydonian himself, but Croydon has given the world a slew of sharp-minded inventors. The Victorian William Stanley was particularly prolific, his list of patents including steel-wheel spider spokes for push bikes, and the all-important 'press for rending steaks tender'. Then there was Kenelm Foss, an astute Croydonian who pioneered the precursor to Pret à Manger. Opened on Oxendon Street in Central London in 1925, Sandy's was the country's first eat-in or takeaway bar of its kind, serving fillings including kedgeree and pheasant. It was a favourite of socialites, Hollywood stars and political big bugs. But most children (and a lot of adults too) would argue that the greatest Croydon creation came from the mind of one Hilary Fisher Page from Sanderstead in South Croydon.

In 1940, Page patented a clever little toy called the Interlocking Building Cube. After the war he refined this into his Kiddicraft Self-Locking Building Bricks, colourful injection-moulded blocks which clicked satisfyingly into place with other such pieces. By the late 1940s the Danish toymakers Ole Kirk Kristiansen and his son Godtfred Kirk got wind of Page's creation and decided to make their own version, called Automatic Binding Bricks. They later contacted Kiddicraft to check that it was OK to do this; Kiddicraft replied that there hadn't been much appetite for their plastic bricks and they hoped Ole and Godtfred had better luck. They certainly did. The Danish company, as you've doubtless worked out, was called Lego.

Following the collapse of the atmospheric railway, Croydon finally settled on steam trains, and made a good fist of it too. In July 1846 the London and Croydon Railway joined forces with the London and Brighton Railway, forming the London, Brighton and South Coast Railway. The Wimbledon and Croydon Railway came along in 1855, and just a year later Purley was joined up with Caterham. All rails, it seemed, led to Croydon. By 1868 the parish of Croydon had 10 stations, linked to four London termini. 'Never, indeed', exclaimed the *Engineering Journal*, 'was a town of the same size more be-railwayed.'

Not everything went according to plan. Central Croydon Station was built on the site of today's town hall as an add-on to what is now East Croydon Station, the idea being that passengers arriving at East Croydon would change for an extra train that would then take them *ever so slightly further* into the town centre. Central Croydon Station opened in 1868, was declared a disaster and closed in 1871. For some reason or other it then reopened in 1886, was declared a disaster a second time, and was put to bed for good in 1890. The London, Brighton and South Coast Railway had overestimated just how lazy its passengers had become. In 1897 the Croydon Camera Club delighted in watching a film of the demolition of the short-lived station – another early demonstration of Croydonians' rubblethirsty obsession with knocking things down.

Still, thanks to thousands of visitors now being funnelled in by train, Croydon was cashing in. 'There is scarcely any town in England,' wrote the *Croydon Chronicle and East Surrey Advertiser* in 1861, 'so well provided with Railway accommodation as Croydon.' Rickety coaching taverns were switched out for the likes of the Railway Hotel in East Croydon, known for its chops and steaks from the grill, dished up in a 'handsome dining saloon'. Croydonians made the most of the new services headed to the seaside. Writes Muriel V. Searle in *Down the Line to Brighton*, 'At Croydon, crowds pushed and fought into the trains, determined to squeeze six-penn'orth of rowdy fun from every penny spent on fares.' Trains ran with up to 44 coaches stuffed with *4,000* passengers – an insane number when you discover that the average Thameslink train has a capacity of just over a quarter of that. Searle also writes about an inspired typo made by one London newspaper, which reported that a train speeding between London and Brighton had struck a cow on the line at Croydon. The cow, said the paper, had been 'cut into calves'. (They meant halves.)

The railway boom was having another effect on Croydon: people were no longer just passing through – they were *moving* there. The census of 1801 records 5,743 Croydonians. By 1837, just before the town had its first proper railway, Croydon's population was 15,000. In the year of Queen Victoria's death, 1901, this number had ballooned to 134,037. Croydon was by now the largest town in Surrey. No longer just a corridor for passers-through, it had become thriving suburbia, a pretty purlieu of London glinting in the peppy new age of the commute. It was now possible to live in this agreeable Surrey market town, catch the train to your job in the City and be back again in time for tea. Croydon had flourished into a dormitory town – a sylvan suburb of London out in leafy Surrey.

Naturally, it didn't take long for people to start complaining about the rail fares. In 1883 the *Sussex Agricultural Express* recorded one Councillor Steele deploring that

> He could remember a time when he could get a return ticket from Croydon to London Bridge for 10d., and he should have thought the Brighton Company would have given them the benefit of the increased population; but instead of reducing the fare, it had been increased from 10d. to 1s. 3d. (shame).

Still, for the first time, the well-to-do gingerly put a toe into Croydon en masse. 'Handsome villas spring up on every side tenanted by City men whose portly persons crowd the trains,' wrote *London Society Magazine* in 1862. An ad in the *Croydon Chronicle and East Surrey Advertiser* from 1903 details a Park Hill villa called High Towers close to East Croydon Station with

> Nine bed chambers, dressing room, two bathrooms, linen room, drawing room, dining room, billard room, morning room, servants' hall, and domestic offices; large gardens, laid out as tennis lawn, artistic flower beds and borders, gravelled walks, kitchen and fruit gardens, detached stabling for five horses, double coach-house, harness room, three living rooms, loft &c.*

This growth spurt brought on by the railways also prompted unwelcome consequences. Overcrowding became a pressing problem, particularly in

---

* It's the '&c.' that really gets you.

Croydon's Old Town. In 1888, one newspaper despaired of Surrey Street as a 'human moral piggery that, for low depravity, either Newcastle or Manchester might match, but certainly could not surpass'. John Ollis Pelton described the area bordered by Surrey Street, Church Street and the High Street as 'neglected and squalid, the playground of dirty children, the lounge of idle and disreputable men and women'. One doctor at the time went as far as calling it 'a kind of cancer – a malignant sore'. Prostitution was rife, and the neighbourhood became known as 'the Disreputable Triangle'. Living conditions were compounded by Croydon's long-running history of flooding.

The Bourne waters, streams that surface intermittently at Caterham and Coulsdon, roughly tracing the routes of the A22 and A23, were a major headache for the Victorians. Liable to inundate Croydon's low-lying Old Town until it resembled a sink of dirty washing-up water, they were nicknamed the 'woe waters', and had been one of the reasons the archbishops jumped ship to Addington. Now in the mid-nineteenth century, the influx of newly minted Croydonians found themselves living in a regular Waterworld: forced to cross roads using planks of wood. In *A Short Chronicle Concerning the Parish of Croydon*, John Corbet Anderson wrote about someone who 'saw a servant at the cellar door get into a floating tub, and push herself with a stick across to where the barrel stood, in order to draw the beer'.

Consequences were often less comical. Flooding was sometimes so bad that children drowned. At funerals, coffins had to be pushed down while the earth was shovelled on top of them for fear they'd float right out of the graveyard. Ponds festered with animal carcasses and pollution from the local gasworks. A local solicitor, William Drummond, gave Laud's Pond in the Old Town the stomach-churning nickname 'Typhus Pudding', while Felix Summerly (the pseudonym for Henry Cole, the man credited with inventing the commercial Christmas card), described 'a great black stagnant cesspool, garnished with dead dogs and decomposing vegetables'. The water quality was so wretched that when potatoes were boiled, they turned black.

'According to an old prophecy', writes Brian Lancaster in *The Croydon Case: Dirty Old Town to Model Town*, 'when the Bourne rises the people of Croydon may look out for death and pestilence.' Indeed, Croydon was rocked by outbreaks of diphtheria, scarlet fever, typhoid and smallpox. In 1848, a terrible cholera epidemic struck.

Something had to be done, and once again Croydon rolled up its sleeves. A year after the cholera outbreak, in 1849, Croydon launched one of the country's first local boards of health. A medical officer was installed. Part of the Bourne was culverted. Running water was fitted into locals' homes (whether they wanted it or not; some were strongly opposed to such an 'inconvenience'). Privies and cesspools were eliminated. Ponds were drained. Proper sewers were dug. On 11 December 1851 the Archbishop of Canterbury, John Bird Sumner (archbishops were clearly still available if there was a ribbon to be cut), opened the new waterworks building just off Surrey Street.

Not only was the Surrey Street Pumping Station a tremendous bit of engineering, able to draw up to half a million gallons of water a day, as well as heating water for the nearby public baths and washhouses, but it was also a savvy piece of recycling. The pump house building was repurposed from the disused atmospheric railway's pumping station at West Croydon, which had been purchased by the Croydon Board of Health for a mere £250.[*]

All this should have been enough to herald a fresh, healthier era for Croydon. But there was a horrendous hiccup: in 1852 a devastating typhoid case struck. 'Death spreads its wings across the town,' announced the *Morning Advertiser*, as Croydonians throughout the town – including, this time, those in the more agreeable quarters – went down with the deadly

---

[*] It was later extended as a baronial folly with battlements and a turret. From an arched window here, the likeness of Princess Diana gazes out. She is the work of street artist Rich Simmons, who tells me, 'I wanted to create a dystopian princess to go in the tower ... Princess Diana came to my mind instantly. Her story was also one of abandonment, and if this was a fairytale maybe she would have been locked in the tower like Rapunzel, waiting for her own prince charming to arrive.'

fever. Many Croydonians blamed the Croydon Board for tinkering with things, and it would take time for them to trust the hygienic new measures. It's also possibly why, when Croydon was granted its own coat of arms in 1886, the motto *'Sanitate Crescamus'* was chosen. It means 'Let us grow in Health'.

Victorian Croydon continued to flourish nevertheless, thanks to both private and civic improvements, filling the vacuum left by the archbishops. Newspapers cranked up their printing presses – the *Croydon Chronicle* in 1855, the *Croydon Advertiser* in 1869, the *Croydon Times* in 1880 – and, as a sign of what had made them viable in the first place, printed railway timetables on the front page. The town's first public park, on the site of the old racecourse at Duppas Hill, opened in 1865. The Croydon Corporation was established in 1883, and began purchasing more land off the church for public use.

By 1889 Croydon was a County Borough, a title that gave it more autonomy, and the chance to show everyone what it was made of. Soon the Disreputable Triangle was demolished. In 1896 a debonair new redbrick and Portland stone town hall was inaugurated, its clock and bells crafted by Gillett & Johnston. Carved reliefs representing the likes of 'Study', 'Reading' and 'Recreation' (naturally there was a sculpture of John Whitgift too) adorned the edifice. Croydon was positively fizzing, a parvenu of a town to be contended with.

Its charcoal-burning days couldn't have felt further behind.

7

RETAIL MAGNET

It was thanks to the suburban swell brought about by the railways that trams were installed in places like Croydon, but this didn't happen overnight. George Francis Train, a quixotic Bostonian entrepreneur who claimed to be the inspiration for Jules Verne's *Around the World in 80 Days*, introduced street trams to Britain in 1860. Unfortunately, most Brits thought Train's scheme was loopy, and he was sent packing back to the US, where he ran for President of the United States, failing at that too. It was almost two decades later, on 9 October 1879, that Croydon's first tram service started running, a horse-pulled tramcar that rumbled along at six miles per hour between Thornton Heath Pond and central Croydon.

One of the key figures behind the tram system was Jabez Spencer Balfour, who'd later become the first Mayor of Croydon. Balfour's business acumen wouldn't always be so whip-smart. After spearheading a duplicitous building society scheme, he was found guilty of embezzlement, *The Times* decrying him as 'One of the most impudent, heartless business scoundrels on record.' Sensing that his goose was cooked, Balfour did a runner to Argentina, but was extradited and sentenced to 14 years hard labour. He wasn't the last Croydon politician to be accused of dodgy dealings.

CROYDONOPOLIS

It took time for many Croydonians to get on board with the trams. Wealthier folk weren't keen on tramfuls of the 'lower classes' being paraded back and forth in front of their villas, while working Croydonians were incensed that it cost double fare to ride on a Sunday, their only day off. There were kinks in the network, too, which rapidly became a confusing medley of competing companies, some private, some municipal. Where the boundary of the Croydon Corporation met that of London County Council's in Norbury, the two tram systems were separated by a comical 6-inch gap, forcing customers to climb off one tram and get on another. In spite of such niggles, tramways spidered out from Croydon into Tooting, Sutton, Purley and Penge. The 16/18 route got you from Croydon to Embankment in an hour and 10 minutes. South London was better connected by tram than it ever would be by the Tube.

All of which led Frederick Thomas Edridge, a man who was Mayor of Croydon on four separate occasions, to unveil in 1896 London's 'dullest plaque'. As uninspiring as the widening of the High Street sounds – and as magnificently metaphoric for those who now equate the town with *everything* bromidic – for Croydon it marked an epochal moment. Not only did it mean trams could now fit along the road, but the scheme also involved the construction of a brand-new parade of flashy shops that would boost the town's profile.

Trams and commercialism went hand in hand. Look at most pictures of early trams, and they're invariably slathered in advertisements. Billboards on wheels, they blaze with seductive brands like Military Pickle, Sunlight Soap, Davies Pea Fed Bacon, and Keen's Mustard.[*] Trams tempted you into town on a shopping spree, and were kind enough to offer you a lift in. And while London's glittering West End was now within easier reach of Croydon than ever, Croydon was not in the mood to be some suburban wallflower coyly twiddling its hair on the sidelines. It was intent on becoming Surrey's answer to Oxford Street.

---

[*] The latter was the product of a successful entrepreneur who lived in Croydon, but sadly is not the origin of the phrase 'keen as mustard'.

RETAIL MAGNET

By the late nineteenth century, the main thoroughfares of Croydon's North End, High Street, George Street and London Road had been transformed into a dizzying array of modern shops. Traditionally, Croydon was known for producing the 'three Bs' – beer, boots and bells. But suddenly its resplendent new avenues of redbricks, their edifices embellished with cupolas, gold mosaic friezes and Flemish gables, offered every necessity, frivolity and exotica you could dream of. In *fin de siècle* Croydon clockmakers ticked, drapers snipped, piano shops tinkled. The air was filled with recherché new aromas, like the freshly-ground coffee dancing out of Wilson's Oriental Café, and a perfume sold by W. Guy Padwick which had notes of white heliotrope, rose and wood violet, and was called A Reminisce of Croydon. The sheer number of shops, and the eclecticism of their wares, were something to behold. A department store called Remsbery's bragged that it stocked everything from wine glasses to horses.

The 1891 brochure *Where to Buy at Croydon* went further, suggesting the *whole world* had set itself at the feet of this on-the-make market town: walk into the Japanese Art Stores, with its elaborate Eastern-styled frontage, and you might come out with a bamboo screen or some carved ivories under your arm. The Indian Cigar Stores would fix you up with a fine smoke, and there was an American dentist who'd happily contour your 'useless teeth' with pure gold. Drink merchant Mr D. H. Weston sold a novel German beer known as lager, 'which is an exceedingly delicate beverage, and may be enjoyed by those with whom English Ales disagree'.

Nipping into town on a Saturday was now a national pastime, and Croydon was one of the finest places to indulge in it. In *The Tramways of South London and Croydon*, the Croydonian historian R. S. Hunt recalls trams crammed with 'bargain-hunting female shoppers', one misfortunate enough to encounter a run-in with the young Hunt's pet rat, Chippy, 'letting forth a shriek like a steam siren' and making a run for the rear door, followed by a flurry of other terrified passengers.

CROYDONOPOLIS

As customers poured into the high streets and shopping arcades, retail magnates began eyeing up the town as a plum investment. In 1882, John James Sainsbury established the first suburban branch of his grocery store on London Road, opposite what is now West Croydon Station. Sainsbury personally oversaw the decor, decking out the shop in fashionable brown and green tiles, Italian marble countertops and stained-glass windows populated with pheasants and hares. '1,000 Canadian Hams', boasted the adverts, 'Only house in Surrey where you can get ... Neufchâtel, Port Du Salut, Green Cheese, Roquefort, Alpine Cream, Camemberts.'

The dime-store tycoon Frank Woolworth sniffed out the opportunities to be had in Croydon too, praising it as a 'fine, progressive and bustling' place for business. The enterprising New Yorker personally opened Woolworths' North End store in 1912, and soon afterwards Annie Venables had the honour of being one of its first shoplifters, charged with pocketing two singlets, two skeins of wool and seven handkerchiefs. (The records say nothing about the theft of any foam bananas from the Pick 'n' Mix, though.)

In a bid to appear as *au fait* as possible, Croydon's shopping scene took on Parisian affectations. Campart's on the High Street sold furs, ostrich feathers and 'Superior Paris Style Hats', a sign in the window reading '*Ici on parle francais*'. Elsewhere, shoppers bought bonnets made 'in the Paris mode' by Miss M. Chalmers Dressmaker and Milliner, while 'tailors, outfitters, hatters, and juvenile clothiers' Staveley Brothers advertised branches in Paris, London, Croydon and Peckham – making it all but impossible nowadays not to think of *Only Fools and Horses*.

Gourmand Parisians themselves might've been tempted by Croydon's foodie scene. At Henry Hopper on George Street there were elaborate menus of 'soup, dressed fish, entrées ... jellies, ices, ice puddings and aerated waters', plus boar's heads and Parisienne biscuits at Christmas. Customers gorged on oysters, jugged hare and roast duck while glugging Californian wines in the palm-studded opulence of the Café Royal (it also did musical

The first suburban Sainsbury's opened in Croydon in 1882, heralding the start of a long relationship. *The Sainsbury Archive*

afternoon teas and catered for 'whist parties, smoking concerts, masonic and other functions').

One of the ritziest dining rooms in town was to be found at Kennards' department store in North End. Described as 'the best appointed anywhere in London – West End or suburbs', it hosted grand luncheons and teas, while the Sparklettes danced or Daphne 'the Golden Voice' Kelf, aka the 'Singing Mannequin', was accompanied by the restaurant's in-house orchestra. Elsewhere in Kennards you could giggle at Punch and Judy shows while taking afternoon tea in the Tudor Restaurant, scoff peach melbas at its Sundae Bar or take your own slice of cake at the 'serve yourself' café, over a century before Croydonians were grabbing wedges of Daim cake from the fridges at IKEA.

Kennards was the first of three major department stores that formed the glitzy kernel of Croydon's shopping Shangri-la. Opened in 1853 by an industrious draper, William Kennard (a promotional booklet from the 1920s described him as a 'manager-cum-buyer-cum-counting-house-clerk-cum-floor-sweeper-cum-window-dresser and sometimes dispatch clerk'), Kennards sold a typically diverse list of goods – perfume, furniture, bicycles, foreign objets d'art, dog food – and prided itself on its customer service. It was somewhere to savour like a fine wine, not just nip in and out of. 'How many times', read one Kennards pamphlet, 'have you hastened your choice and bought something you did not *really* like simply because an importunate and misguided assistant stood by your side and indicated as plainly as possible that she wished you to hurry up? That sort of thing does not happen at Kennards.'

Then in 1921 an Australian wunderkind by the name of Robert 'Jimmy' Driscoll was enlisted, and Kennards went into overdrive. Driscoll, a bona fide firebrand – suddenly with one of the country's most formidable retail kingdoms at his fingertips – set about blurring the line between department store and theme park. He was part Gordon Selfridge, part Walt Disney.

Driscoll had dreamed up the slogan 'We Entertain to Sell', and he meant it. On his watch, Kennards began hosting fashion parades, dog shows, beautiful baby competitions, swimsuit pageants and palm readers. A 'playground in the sky' offered an outdoor theatre and a tennis court, while a two-level golf course featured one fairway indoors, the other perched on the rooftop. Real-life cheetahs were plonked on dinner tables in the restaurant. Perhaps inspired by Selfridges' 1909 publicity stunt in which Louis Blériot's monoplane was displayed on the shop floor, Driscoll bagged not one but two of Malcolm Campbell's record-breaking *Blue Bird* cars, showing them off in the shop in 1931 and 1933. Ever the showman, Driscoll sat behind the wheel posing for the cameras.

Driscoll also understood the importance of keeping children distracted while parents shopped. The book *Kennards of Croydon: The Store That Entertained to Sell* tells how kids rode around on a miniature railway on the department store's rooftop or stood in dumbstruck wonderment at special guests like the Great Omi, a man covered from head to toe in tattoos, and Max, Moritz and Akka, three roller-skating chimps in top hats and tails.

There were Shetland pony rides through the arcade (a tradition that lasted up until 1966), and for a brief time a zoo. ('The cages are adequately strong,' ran a reassuring disclaimer, 'but we cannot accept responsibility for what may happen if people will poke their fingers through the bars.') Kennards had regular installations like its 'Abyssinian' and 'Indian' villages while, with

Crowds enjoy the rooftop Punch and Judy show at Kennards, the Croydon department store that boasted it 'entertained to sell'.

CROYDONOPOLIS

another wave of the Disneyesque wand, an 'Elizabethan Tudor Arcade' was opened in the 1950s, its ersatz half-timber shopfronts, cobbled streets and 'Bun Shoppe' at odds with the fact that, by the end of the decade, Croydon would be tearing its architectural heritage limb-from-limb.

A-lister celebrities saw Kennards as a chance for self-promotion. In the post-war years, Dirk Bogarde, Anna May Wong and Norman Wisdom waltzed in through its doors. The entertainer Googie Withers came in 1951, cutting the cake celebrating the store's upcoming centenary and predicting that 'in another 100 years another actress will be saying just what I say today, but she will travel here in five seconds by rocket'). Another star to grace Kennards was Scruffy, a superstar pooch who'd acted alongside Henry Fonda, Vivien Leigh and Rex Harrison, and had even published his own bestselling autobiography.

Animals were good for drumming up business. *London's Lost Department Stores* records Jimmy Driscoll promoting a 'Jumbo Sale' by borrowing two elephants from a Bertram Mills' circus, blocking both ends of the main street, and earning himself a court appearance. Another escapade saw Joss the baboon flee his cage in Kennards' Monkey Village, go on a rampage through the toy department, then clamber up onto the roof where, 'to the delight of everyone watching him in North End, he danced.' All publicity was good publicity, and for three decades under Driscoll the tills in Kennards rang out and the Lamson Pneumatic Tube System went into overdrive, sending customers' cash flying through pneumatic pipes at 30 miles an hour to an Automatic Central Desk, before their change and a receipt were fired right back at them.

But Kennards had stiff competition. Two other department stores, Allders (which opened in 1862) and Grants (1894), were in hot pursuit. Grants came to be regarded as the swishest of the triumvirate, what with its fur department, travel and theatre booking bureau and food hall with a frozen section offering 'strawberries from the Kentish beds, apricots and melons, cherries still glinting with last year's sunshine'. Like Kennards, Grants had an upmarket restaurant, although theirs was frequented by dashing pilots from the nearby airport. One regular was Gordon Olley, who'd battled it out with the Red Baron back in the day and, according to *Croydon Airport: The Great Days*, was often seen dining at Grants in a Savile Row suit, smoking Craven 'A' cigarettes.

Paving the way for twenty-first- century London's fad for rooftop bars, Grants offered customers the chance to sink cocktails amid dahlia beds,

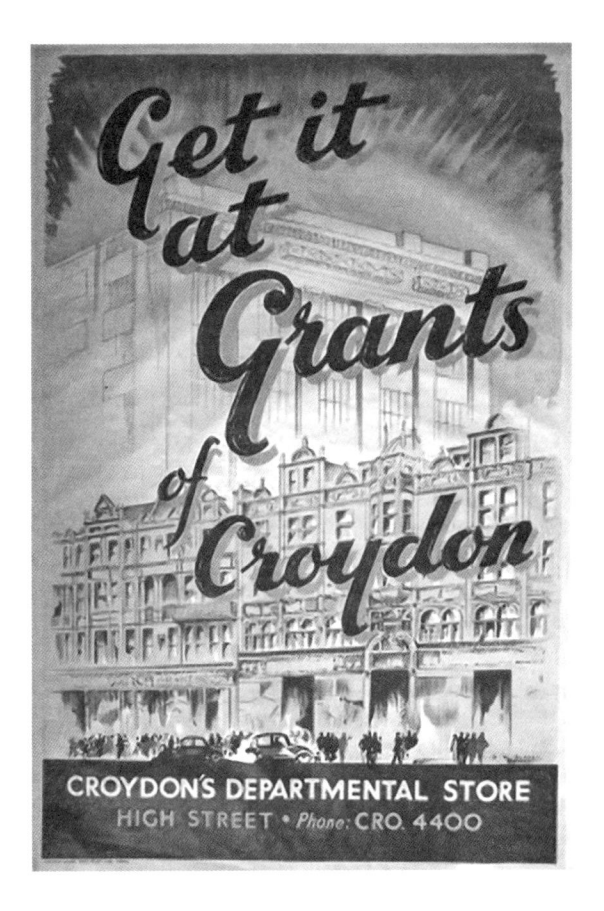

Such was Grants' reputation during its interwar height that French aristocrats flew to Croydon just to buy their suits here.

watching the planes take off and land. Even come the 1960s, Grants was swanky enough to host Elizabeth II and her husband Philip, who visited for afternoon tea. Allders, on the other hand, was renowned for its family mourning department: 'Special attention given', it guaranteed. There was nothing you couldn't buy in the Croydon of that era – even a tasteful send-off for a recently departed loved one.

The three-way rivalry meant Christmases in Croydon were legendary, the big three jostling to outdo one another. An arsenal of Santas vied for the kids' attention. The humble sleigh and reindeer were soon superseded as too pedestrian, and Santa's chosen methods of transport flitted from miniature car to aeroplane to helicopter to the Kennards 'Dreamland Express'. In 1938, kids were invited to visit the Man in Red aboard the 'SS *Santa*', a treasure ship that had 'run aground on a South Sea Island Reef'. In 1943, in the midst of wartime rationing, Kennards gloated it had 'the largest Father Christmas in the World'. Massive toy fairs were laid on, and kids quickly learned how to plead and pester for the contraption of their dreams.

The adults weren't much better, especially thanks to the rise of the 'unmissable deal'. Kennards had both 'Blue Pencil' week (another Driscoll brainchild, and one which later morphed into its successor Debenhams' famous Blue Cross Sale), and 'Clock Days', on which a new bargain was announced every 60 minutes. Tessa Boase, author of *London's Lost Department Stores*, imagines being one of those avid bargain-hunters: 'I'd go into Kennards during one of its "one-hour" sales, chasing around the departments, frantically grabbing whatever was being discounted as the clock struck the hour. It's the electricity-mad Twenties: I want to see what "electric servants" are on offer –waffle iron, toaster oven, washing machine . . .'

Given all its other technical experiments in transportation, it's unsurprising that Croydon was also a testing ground for trams powered by battery, internal combustion engine and gas – all unsuccessful, and the latter giving

RETAIL MAGNET

Allders, opened in 1862, was part of Croydon's triumvirate of sumptuous department stores and, by the new millennium, the third-largest department store in the entire country.

passengers terrible headaches and nausea. Still, just after the turn of the century Croydon was one of the first parts of the Greater London area to have electric trams: 'This can't be Croydon...' gasped the *Croydon Advertiser* as a parade of 20 electric tramcars purred through the streets during the launch of the system on 26 September 1901. 'To see huge cars going along driven and lit by electricity as light as day inside at night was a sight an old inhabitant must have marvelled at and even the new inhabitant must have wondered about.'

This *was* Croydon, though, and the age of the tram had truly arrived.

In the first half of 1902, Croydon's tram network carried 3.69 million passengers – over 27 times the entire population of Croydon. It was, as Robert J. Harley put it in *Croydon Tramways*, 'tramwaymania'. Even the habitual Croydonophobe David Lean was so ensorcelled by the 'exciting flash and crackle' of the tram's arm zipping along the wire that he would put one in his adaptation of *Doctor Zhivago*.

By the 1930s, though, a handful of wealthy shoppers were arriving into town using an altogether different method of transport. 'Grants was so high-end,' Tessa Boase tells me, 'that French aristocrats would fly into Croydon Airport simply to be measured for a bespoke suit.'

The age of the aeroplane had taken off.

8

TAKING WING

The frothing mass of 120,000 fans waited impatiently for their American idol to appear. 'Excitement rose to fever heat, barriers and police failed to hold back the crowd,' reported a *British Pathé* reel. As the field beneath 240,000 surging feet was churned into mud, security realised they had a problem. Because now, as the man of the moment emerged from the clouds to peer down on Croydon Aerodrome, he could see there was no space for him to land. If he wasn't careful, he was going to flatten a field full of hysterical spectators.

The twenty-ninth of May 1927, and Charles Lindbergh, who had just completed his landmark solo transatlantic flight from New York to Paris non-stop, was now tracking back to Croydon to soak up the plaudits. But as his *Spirit of St Louis* monoplane circled over the aerodrome, the audacious Michigander grew more frustrated by the second. What was *going on* down there? The chaos had only just begun. As Lindbergh circled round and round, and the police continued to flail and grapple with the multitudes below, a group of 50 onlookers narrowly escaped tragedy when the roof of an outbuilding they'd shimmied up onto collapsed. To add to the fracas, a passenger plane nose-dived into the ground. It was farcical. The crowd, later grumbled the editor of the *Aeroplane*, had 'behaved just like a lot of foreigners'. In their defence, some hadn't seen an aeroplane until now, far fewer one flown by one of the world's most famous men. At the height of Beatlemania in 1964, when the Fab Four landed back at Heathrow after their tour of the States, they were greeted by 12,000 fans. Lindbergh's appearance at Croydon had summoned up *10 times* as many.

After wheeling over Croydon for 10 or so minutes, Lindbergh finally identified a strip large enough to touch down on. The moment it landed, the *Spirit of St Louis* was mobbed, and its pilot could only count himself lucky the plane wasn't torn wing from wing by frenzied souvenir hunters. On subsequent visits to Croydon Lindbergh wore a cloth cap and dark glasses as a disguise.

The golden era of flight was in its ascendency, and Croydon was smack dab in the heart of it. But *why?*

In 1911 the Croydon filmmakers Cricks and Martin released *The Pirates of 1920*. In this short film, a motley band of steampunk thugs cruise through the skies in a rugby ball-shaped airship, dropping cartoonish bombs onto ships and clambering down rope ladders to abduct women. It played on a fear that was beginning to give people in the real world the jitters, and justifiably so. Just four years after *The Pirates of 1920* was released, aerial assailants started raining down fire and brimstone on Croydon, and this wasn't make-believe.

Airships weren't an altogether alien concept to Croydonians. Since the early nineteenth century Crystal Palace Park had served as a launchpad for balloon flights – some successful, some toppling the local chimney pots, some plummeting to earth with fatal consequences. By the beginning of the next century these balloons had evolved into steerable, 'dirigible' craft. Some Croydonians would even have glimpsed the first airship ever to launch in Britain, as it rose from Crystal Palace on 19 September 1902 with the

When Charles Lindbergh flew his *Spirit of St Louis* into Croydon in 1924, some 120,000 spectators showed up to greet him. *Alamy*

moustachioed aeronaut Stanley Spencer at the helm. While Spencer's craft, which had an advert for Mellin's baby food plastered on the side, drifted over Streatham, then Clapham, towards Chelsea and on to Harrow, he lobbed rubber balls over the side. This wasn't some practical joke: Spencer was demonstrating the damage that might be wreaked by airborne invaders. Soon enough, both Spencer's rubber ball stunt and the plot of *The Pirates of 1920* would ring truer than anyone's deepest fears had allowed them to imagine.

In February 1913 there were reports of a 'strange airship' seen drifting over Croydon. It was soon identified as the German craft *Hansa*. Britain was on the cusp of war with Germany, but while fighting was to be kept mostly to the trenches, already there were indications that war could play out in the skies, too. 'ZEPPELIN RAIDS TO BE EXPECTED FROM "MAD MEN"', splashed a *Croydon Times* headline, not exactly worded to calm anyone's nerves. Hysteria set in. A West Croydon woman found herself in the dock after causing a large crowd to form on George Street: 'Look! There's a Zeppelin!' she'd cried, pointing to the heavens, and when a police officer told her what she could see was in fact the Moon, and to kindly move on, she'd responded 'using very filthy language'. Businesses realised there was money to be made. 'Zeppelin raid may be in Croydon any night,' warned one advert for a shop on London Road; 'every man, woman and child should buy an army respirator and protect yourselves from poisonous gases – 6½d each.' Norman and Co. of Katherine Street ran an ad forewarning of 'unexpected visitors' and urging Croydonians to take out Zeppelin bomb insurance on their homes: 'These risks are NOT covered by normal Fire Policies.' An invasion of Croydon now felt only a matter of time.

In the early evening of 13 October 1915, a squadron of five German airships appeared on the Norfolk coast and slipped across the skies towards London. Navigation was primitive, and only one of the crafts made it to the West End, the others causing havoc wherever they could. For the wayward L14 Zeppelin, commandeered by Kapitänleutnant Alois Böcker, this turned out to be East Croydon. 'A flash from the sky, a sudden illumination of the whole neighbourhood, a deafening explosion and violent tremors of the ground showed that the German invaders had actually reached Croydon,' recalled one observer later. Air raid sirens weren't a thing at this time; it was thought these would only bring people out onto the streets to gawp at the 'baby killers' working their terrible magic, and lead to more casualties. Between 14 and 18 bombs were dropped on Croydon that night. The first hit two houses on Edridge Road where, according to *Croydon and the Great War*, 'a mother and daughter who were in bed in one of them were thrown, bedstead and all, into the street. A baby boy in the other was pinned down by a falling roof but miraculously escaped injury.' Not everyone was so fortunate. A second bomb fell on 12 Beech House Road, where a man, housekeeper and three brothers, Brien, Roy and Gordon Currie (aged 10, 14 and 15 respectively), were sleeping. Roy was killed instantly, Brien passed away on arrival at the hospital and Gordon died soon after from shock. Croydonians on Oval Road and Stretton Road were killed too – in all, nine perished in that haphazard airship raid. The British authorities now knew they had to act fast.

Consequently, 407 acres of Manor Farm in Wallington about two miles south-west of central Croydon were requisitioned by the Admiralty, who tacked up a series of canvas and timber Bessonneau hangars and populated them with a couple of BE.2 biplanes, plus two pilots to fly them. The pilots slept next to their planes, ready at any given moment to be woken by a call from the War Office ordering them to take to the skies. And just like that, the Government Air Operations Field Beddington, aka Beddington Aerodrome, was born. Along with a series of other hastily built airfields orbiting London, this formed the capital's ring of defence. Though no one realised it at the time, Croydon's airfield, however, was to prove altogether special.

By the tail end of the First World War, aircraft were playing an increasingly salient role, and so in January 1918 the National Aircraft Factory No. 1 opened in Waddon, just down the road from the aerodrome at Beddington. This huge set-up – 58 buildings across 240 acres – was tasked with churning out 200 aeroplanes a month to feed the insatiable war machine. A second airfield was at Waddon itself, purely to test-fly the new Airco DH.9 biplane bombers now being hurried off the production line.

But exactly one year after opening, the Allies won the war, and what had been a factory for assembling planes became one for dismantling them. Fifteen hundred workers found themselves out of a job overnight. It looked like curtains for both of Croydon's embryonic aerodromes.

And then fate struck. In 1919, Hounslow Heath Aerodrome in North

CROYDONOPOLIS

London had started running the British Empire's first scheduled daily international commercial flights. Yet less than a year later, it was stripped of that role and returned to the military. Britain's first proper international aerodrome, the brass hats had decided, wasn't going to be in Hounslow, but Croydon. The two airfields at Beddington and Waddon pooled their resources. On 29 March 1920, Croydon Aerodrome[*] officially opened as a customs airport. It might just have been the most significant date in Croydon's history since John Whitgift had rocked up back in 1583.

The creation of the airport was improvised and messy. Just as William Goldman once wrote of Hollywood that 'nobody knows anything', so in the early days of air travel no-one really knew what they were doing either. To start with, the pilots who'd been relocated from Hounslow to Croydon didn't even know how to *get* there. In *The Seven Skies*, his history of British civil aviation, John Pudney writes of how 'the first little party of pioneer airwaymen to invade Croydon, driving in an old Ford car, lost themselves in a maze of roads between Sutton and Wallington ... and it was some time before they located the aerodrome.' When they finally found it, sheep had to be shooed off the runway. Arrivals and departures were chalked up on two boards resembling pub menus. The control tower was little more than a wooden hut on stilts accessed by a ladder, inside which one visiting journalist described 'magicians playing about with little levers and handles'. The customs house was a glorified barn, with signs tacked up above two doors: 'British' and 'Non British'. The whole ramshackle assortment of buildings, commented Robert Brenard, Imperial Airways' first PR man, 'reminded one forcibly of a Wild West township'. Cars driving along Plough Lane were regularly stopped by a man waving a red flag, allowing planes to taxi across from hangar to runway, towed by a tractor. Pilots figured out their position by dipping down

---

[*] Though it was officially Croydon Aerodrome, many stubbornly continued to call it Waddon Aerodrome. A nickname the papers favoured was the 'Charing Cross of the air', a nod to the Central London railway terminus.

over railway stations to read the name board and, when they were looking to land, sniffed the air for the ripe aroma of the nearby Beddington sewage works. Plane cockpits were invariably open, and anyone who went up in the clouds came back down sitting in pools of water. The birth of international air travel was crude, but Croydon was undeniably the cradle. As if to hammer this home, its name was daubed in stentorian white, chalky capital letters by the runway: 'CROYDON', it yelled up into the heavens.

Between the wars Croydon saw prime ministers fly in and out for important rendezvous: Ramsey MacDonald, in his open cockpit, sporting goggles and leathers – a dashing look for a serving prime minister. Anthony Eden preferred to dig into a detective novel while aloft. As an international gateway, Croydon was soon playing host to royalty from all corners of the globe. In March 1928 crowds gathered to watch Afghanistan's King

At the original, somewhat cobbled-together Croydon Aerodrome, seen here in 1922, sheep had to be shooed off the runway. Imperial Airways' first PR man was reminded 'forcibly of a Wild West township'. *Mary Evans*

Amanullah take off on an aerial sightseeing tour, which passed over the Tower of London, Buckingham Palace and the Crystal Palace. Among the great and the good of the entertainment world who would step on or off a plane at Croydon were Josephine Baker (majestically swaddled in furs), Babe Ruth, Fred Astaire, John F. Kennedy, Hollywood couple Mary Pickford and Douglas Fairbanks, Indian Chief Eagle Elk of the Sioux tribe (who posed for a photograph beside a plane with some of his warriors in headdresses), Rita Hayworth (admittedly owing to a forced landing in bad weather) and Charlie Chaplin.

While there's an infamous tale about the kidnapping of Charlie Chaplin's corpse by graverobbers in 1978, Chaplin was also briefly abducted while he was alive and well – and it happened at Croydon. In 1921 the film star was riding high. He'd just released his latest masterpiece, *The Kid*, and was due to meet the Prime Minister, David Lloyd George, while over in Europe. Chaplin had previously promised the film director, producer and owner of Clapham's Majestic cinema, Castleton Knight, that he'd come and visit the picture house while he was in town.

With this promise now looking unlikely to be upheld, Knight took matters into his own hands, dressing up as a chauffeur (complete with pasted-on moustache), pulling up in a car as Chaplin landed at Croydon and announcing that he was Chaplin's ride to the Savoy. 'I let the car rip along Mitcham Common,' bragged Knight in a subsequent article in *Kinematograph Weekly*, 'and then in order to get to Clapham without causing suspicion I drove a roundabout way …' Assuring Chaplin and his secretary that they were indeed headed for the Savoy, Knight eventually pulled the car up outside his cinema instead, feigning engine trouble, before exclaiming, 'I hope you will not mind, Mr Chaplin, but I've kidnapped you!' With this, he ripped off his fake moustache and cap, declaring with a hammy flourish, 'I'm Castleton Knight and this is the theatre you promised to visit!' Fortunately, Knight had caught Chaplin in a sanguine mood; the star cracked up laughing, and since one of his films happened to be playing at the time, it was paused so the man himself could make an impromptu speech to an astounded audience.

Modern-day airports have security up to the eyeballs, but Croydon Aerodrome actively beckoned sightseers inside. Full-time tour guides proudly showed visitors around the landmarks: the hotel, post office (from which souvenir postcards could be sent), tea gardens and palm court. Most visitors weren't here to catch a flight, but a glimpse of the pilots. In their sheepskin-lined leather jackets, flying caps and goggles they were characters straight out of an adventure book – Biggles in the flesh. If Croydon Aerodrome was the Wild West, then here were the cowboys. They knew no fear, and had names like 'Dizzy', 'Count Vodka' and Scruffy 'the Undertaker' Robinson.* One pilot, Ray Hinchcliffe, wore a patch over one eye, and had a habit of sliding a bowler hat on his head the moment he'd landed. A Belgian aviator caused a stir while wheeling around Croydon Aerodrome, thanks to his provocative plane registration: 'O-B.A.B.Y.' These men were plainly sex on wings.

A chap called F. L. Barnard became the inaugural winner of the King's Cup air race. First held on 8 September 1922, this 810-mile grand prix of the skies started and ended at Croydon, 22 British Empire planes nipping from post to post in Birmingham, Newcastle, Glasgow, Manchester and Bristol, then back up to Croydon – real *Dastardly and Muttley in Their Flying Machines* stuff. The feverish betting of Croydon's horse racing days briefly returned: 'It was odd to find a bookmaker so enthusiastic,' wrote one journalist, 'as to be calling out the odds at half-past five in the morning to the crowd of spectators that had assembled.'

While the pilots were battling it out at 150 miles per hour elsewhere, the

---

* Robinson apparently 'looked like a clown and flew like an angel', although he earned the 'Undertaker' bit of his name by crashing into a cemetery taking off from Ostend, so clearly didn't fly like an angel all the time.

Croydon crowds were kept entertained with feats of aerobatics, parachute jumps and the phenomenon of 'balloon sniping', in which large balloons were let up into the sky every 30 seconds, followed by men in planes brandishing shotguns to shoot them down against the clock. After Barnard raced across the finish line in his blue and silver *City of York* (the same plane he and his wife had flown to Scotland in for their honeymoon), camera crews and pressmen hustled the bone-weary victor for a quote. He gave them one: 'I want a bath.'

Thrilling though it was, the King's Cup was merely a metaphorical rev of the engine. The race was now on to get the furthest, fastest – and Croydon was in the cockpit. Alan Cobham became the first person to fly from England to Cape Town and back again. Returning to Croydon in 1926 with a triumphant escort of little Moths and two big airliners, he was greeted by a sea of fluttering handkerchiefs. In March 1930 R. N. Chawla and Apsy Engineer became the first Indians to fly from India to England, successfully landing at Croydon in a de Havilland Gipsy Moth with no radio. Engineer was just 17. In 1931 the Scot Jim Mollison flew from Australia to Britain in a record-breaking eight days and 19 hours and, much to the bemusement of the knackered pilot, was greeted at Croydon by a real-life kangaroo squaring up to him.

Aussies of the human variety were common visitors to Croydon. Brisbane's Charles 'Smithy' Kingsford Smith whisked the first official air mail from Australia to Croydon in 1931, while Bert Hinkler, the 'Australian Lone Eagle', flew the first solo flight from England to Australia in 1928, setting off from Croydon on 7 February, landing in Darwin less than 16 days later, and becoming an instant pin-up. There's a wonderful caricature of Hinkler leaping into the air, his head superimposed on the body of a kangaroo, drawn by the Croydon-based artist, Charles C. Dickson.

Bewitched by the aerodrome from its earliest incarnation, Dickson smartly became pals with many of the pilots around Croydon, sketching their cartoons in the hope they'd take him up in their planes – which they often did. Many of these illustrations ended up framed in the Pilots' Bar of Croydon's Aerodrome Hotel; the *Strand* magazine published a photo in 1931 showing a huddle of pilots knocking back a beer in front of a gallery of their own caricatures.

While the competition was fierce, the camaraderie, especially with so many of these pilots having just made it through a war, was robust. Emboldened by his success flying from Karachi to Croydon, Apsy Engineer had almost immediately flown back towards India alone, only to find himself stranded in Egypt thanks to a faulty spark plug, for which, in spite of his surname, he had not packed a spare. While he was still wondering how he was going to get out of this scrape, another pilot, J. R. D. Tata, who was flying the route from the other direction, landed in Egypt and lent Engineer his spare spark plug. The pilot Bill Lawford remembered how Sefton Brancker, the Director of Civil Aviation, would often lift spirits at Croydon. 'After any particularly nasty flight', wrote Lawford, 'there was always awaiting his great hearty grip of the hand, with a twinkling "Well done – Cheerio and stick it, old lad."'

Not all the records were being set by the aviators themselves. In 1927, 92-year-old Elizabeth Reeves, dressed, as the *Evening Standard* described her, in 'Victorian period' clothes, which really puts into perspective just how newfangled this aeroplane lark was, went on her first flight, becoming surely the oldest person at that time to ascend into the heavens, and come back down again. 'I really don't feel at all afraid,' Elizabeth had smiled as she stepped onto the plane at Croydon, 'but I had two small nips of whisky before I left home.'

Croydon and fearless female flyers go hand in hand. In an era when women still weren't allowed to drive Croydon's trams, the phenomenon of aviatrixes took hold before most men knew what was happening. In 1930 Winifred Brown, who'd learned to roll cigarettes at five, and been expelled from school at 14 for scrawling 'The headmistress can go to hell' on the toilet wall, became the first woman to win the King's Cup. Another Winifred, Winifred Spooner, crashed into the Tyrrhenian Sea en route from Croydon to Cape Town in 1931 and swam two miles ashore to fetch help for her co-pilot. Spooner wound up crashing so frequently, indeed, she earned the nickname 'Bad Luck Wimpey'. None of these accidents were serious enough to hurt her badly (though one did leave her with tattered stockings), but misfortune hounded Wimpey nevertheless, and she died aged 32 following a short bout of pneumonia. Lady Mary Heath – 'Britain's Lady Lindy', as the Americans dubbed her – was another aviatrix familiar to Croydon. She flew back from Pretoria in South Africa, her luggage including a bible, a shotgun and a tennis racquet (the latter jettisoned over the side as she desperately tried to shed some weight as her plane skimmed an East African mountain range).

Piloting a plane was usually a young woman's game, though it didn't have to be. Mary, Duchess of Bedford went up in her first plane, from Croydon to Woburn, at the age of 60. She'd done so on the advice of her doctors, to combat an incessant buzzing sound in her ears, and soon decided she wanted to get behind the joystick. Two years later, on 9 August 1929, Mary returned to Croydon following a record-breaking 10,000-mile flight to India and back, to a rapturous reception. As a tiny speck in the sky grew into her blue and gold monoplane *The Spider*, 'Hats and handkerchiefs were waved wildly,' the *Scotsman* reported. 'The crowd tore across toward the 'plane, officials' efforts to keep them back being in vain ... When [Mary] stepped out of the machine she was immediately surrounded by a ring of outstretched hands. Fifty cameras clicked at once.'

Not only were the women beating the men to many of the juiciest records but they were also, on the whole, being treated with equal reverence. Not in the habit of missing a trick, Kennards' impresario Jimmy Driscoll invited the Kiwi aviatrix Jean Batten, fresh from becoming the first woman to fly to and from Britain and Australia in 1937, to the department store for a meet-and-greet. 'As she entered the store,' wrote the *Croydon Times*, 'the cry went up, "Bravo, Jean!" followed by a big burst of cheering.'

Not content with proving themselves accomplished flyers, some women turned their hand to being aviating entrepreneurs. Mrs Mildred Mary Bruce made her name in February 1931 when she landed at Croydon having made a 19,000-mile flight (the longest ever made in a light plane) to Japan and back. While in Hanoi she was awarded the Order of the Million Elephants and the White Parasol, which allowed her to call a million elephants to her aid, should she ever feel the need. That same year the 19-year-old Peggy Salaman went one better, returning from a record-breaking flight to Cape Town with two tiger cubs, Juba and Joker, she'd picked up in South Sudan. Pet passports and vaccines, you'd assume, were not yet a thing.

Then there was Amy Johnson. Born in Hull in 1903, the same year the Wright Brothers made their first flight, Johnson was the unassuming daughter of a fish merchant. In 1925 Johnson took her first flight in a plane, with her sister Mollie. 'We both enjoyed it, but I would have liked to have done some stunts,' Johnson wrote to a friend. She was hooked, and a few years later signed up for flying lessons at Stag Lane Aerodrome in Edgware – lessons, as it happens, she took from the comic actor Will Hay, who lived in Croydon.

Johnson wasn't as wealthy as some of her aviatrix counterparts. She paid for her lessons by working as a secretary in a solicitor's office, and babysat at nights. She watched on in envy as Bert Hinkler completed his trip to Australia, vowing there and then to better him. 'I will beat his record, his fifteen-day effort, I am certain of it,' wrote Johnson. 'Why not this twenty-six-year-old woman?'

On the evening of 4 May 1930, Amy Johnson flew her second-hand, radioless Gipsy Moth *Jason* from Stag Lane to Croydon Aerodrome, ready for departure to Australia early the next morning. After a sleepless night at the Aerodrome Hotel – not nerves, of course, but the traffic on the incessant

Purley Way – Johnson, dressed in her fur-collared flying suit, set off. At least, she tried to. 'As Amy tried to gain speed down the runway,' Sally Smith writes in *Magnificent Women and Flying Machines*, 'she realised she wasn't going to make it. She slowed down and turned around for another go.' *Jason*, bottle green with silver-grey wings, was loaded up with a cumbersome amount of fuel and equipment, and it took another attempt to get the craft airborne. As it hauled itself unsteadily into the sky, only a handful of people were there to wave the plane off. There'd been little coverage of Johnson's attempt, and this suited her; she preferred her own company and thought she was probably going to fail anyway. Johnson's self-confidence had been bruised ever since she'd had her front teeth knocked out by a cricket ball at the age of 14. At least up in the heavens now she was alone with nothing but *Jason* and her own thoughts.

To Johnson, her aircraft was more than just a machine. The actor Jenny Lockyer, who has turned Johnson's escapades into a one-person play, *Last Flight Out*, tells me of *Jason*: 'He played a huge part in Amy's life and the way she talks about him, [it's as though she's] referring to him as a dear, dear friend.' Johnson's written account of the flight to Australia often shows her fretting over *Jason*'s health: 'I attempted to climb above this cloud cover but *Jason* sounded so very unwell at 10,000 feet,' she wrote on day three. As news spread that Johnson and her winged companion were on course to reach Australia, excitement bubbled up in Britain. 'Who would have thought that a girl born in a simple house down St George's Road in Hull would ever be newsworthy?' Johnson wrote.

She didn't beat Bert Hinkler's record; in fact, she came in three-and-a-half days over. That didn't matter, though. She was still the first woman to fly solo from Britain to Australia, more than enough to spark an eruption of Amymania. When Johnson flew back into Croydon on an Imperial Airways flight on 4 August 1930, not only were there 100,000 people at the airport to greet her, but a cavalcade of *one million* also lined the 12-mile route from

Amy Johnson, the formidable aviatrix, arrives at Croydon from Japan in her plane *Jason II*, as she did on many occasions. *Alamy*

Croydon to the Grosvenor House Hotel, where a grand reception awaited. Lindbergh could eat his heart out. A pop song, 'Amy, Wonderful Amy', rotated on gramophones across the country: 'Believe me, Amy, you cannot blame me, Amy,' crooned Jack Hylton, over sound effects of a puttering plane engine, 'for falling in love with you.' Women copied her hairstyle – the 'Amy Johnson wave' preceded the 'Rachel-from-*Friends*' by over 60 years.

While women aspired to be her, men fell over each other to be in Johnson's company. Three months after her success Johnson was back in Croydon for a lunch thrown in her honour, attended by J. B. Priestley, Alfred Hitchcock and Noel Coward. At another luncheon in 1931 Charlie Chaplin and George Bernard Shaw were far more interested in talking to her than one another.

But all Amy Johnson was really interested in was feeding her insatiable sense of adventure.

She went on to polish off record after record: Croydon to Moscow in 21 hours in 1931; Croydon to Tokyo and back in record time (also in 1931); a new record from Croydon to Cape Town in 1932. Just as Amelia Earhart flew the flag for women in the United States, Johnson was doing it for the Brits. The two met and became friends. There's a picture of them strolling along the beach in Atlantic City in 1933; Earhart is in pale T-shirt and slacks, Johnson wears a summery playsuit decorated in chevrons: they are the epitome of cool, the most vital women of the moment.

'I am never so happy,' Amy Johnson had said in her victorious Australia speech at Croydon, 'as when alone in the silent spaces of the sky.' Yet two years later she married Jim Mollison, the fiery Scot who'd once almost been punched by a kangaroo. The two met on a commercial flight, and were engaged just *eight hours* later, showing just how impulsive and bloody-minded both could be. According to Mollison, his telephoned proposal went like this:

Jim: 'Would you like to get married sometime?'
Amy: 'Well, Jim, I've never tried it.'
Jim: 'Amy, shall we get married?'
Amy: 'I'll take a chance.'

Together they set off on several escapades, becoming the prototype for the modern-day celebrity couple and yet infinitely more enthralling – all romance, risk and record-breaking bundled into one radical package. It was, giggled the papers, an 'aeromance', dubbing them the 'Flying Sweethearts'. Yet for all the photos of them together, the one that captures Johnson and Mollison at their most candid shows their plane, *Seafarer*, written off on the runway at Croydon. A botched launch in which the plane was overloaded with petrol saw it lurch into a dip in the runway during take-off and get badly damaged. Johnson, in white overalls, looks on pensively. Mollison, arms stubbornly crossed, wears a sour scowl. He'd been on the beers until three that morning and, as some have pointed out, it's probably for the best that *Seafarer* never took flight that day. As Mollison's envy of his successful wife grew (her 1932 Cape Town feat smashed Mollison's record by 10 hours), so did his drinking and philandering, and the Flying Sweethearts became the Flying Divorcees. Amy Johnson had been born to fly solo after all.

Amy Johnson eventually died doing what she loved best,[*] as did so many pilots of the Croydon era. The envelope, after all, was being pushed on a daily basis. Other pre-eminent figures only just escaped the grim reaper. Winston Churchill had taken flying lessons at Beddington Aerodrome, much to the chagrin of his poor instructors. A Mr Toad of the skies – enraptured by flying machines but ham-fisted at the controls – Churchill very nearly got himself killed one summer evening in 1919. 'The aeroplane was just turning from its side-slip into the nose dive,' he recalled later in his memoirs, *Thoughts and Adventures*, 'when it struck the ground at perhaps fifty-miles-an-hour with terrific force ... I felt myself driven forward as if in some new dimension by a frightful and overwhelming force, through a space I could not measure.' The quick thinking of Churchill's instructor, who turned off the ignition to prevent the plane from bursting into flames, saved them both, and Churchill got away with bad bruising. This was one of three self-inflicted plane crashes Churchill had at the aerodrome.

The Prince of Wales (later briefly Edward VIII) had a similarly reckless flying record; his dad George V ended up banning him from Croydon after

---

[*] In January 1941, she was ferrying a twin-engine Airspeed Oxford from Prestwick to Kidlington near Oxford for the RAF when she got lost. Finding herself running out of fuel, she climbed out of the cockpit of the speeding plane onto the wing, and leaped off into the frigid waters of the Thames Estuary. A naval vessel, HMS *Haslemere*, caught sight of her, and as a voice was heard – 'Hurry, please hurry!' – attempts were made to pull her from the water. She was hit by the ship's propellor and never seen again.

he teamed up with a one-armed major and started showing off in aerobatic displays. Edward's eminently more sensible brother (later King George VI) was more trustworthy behind the controls, earning his wings at Beddington Aerodrome with little fanfare. The two brothers' experiences in the air would prove an apt metaphor for their attitudes towards wearing the Crown.

While air travel often proved a risky business for pilots, nervous passengers had their reasons for pre-flight nips of whisky too. 'If one marked with red spots all the fields where airliners of the Twenties and Thirties made forced landings', Charles C. Dickson once quipped, 'the map of Kent would look like a severe case of measles.' In September 1922 hundreds of pleasure-seekers on the seafront at Folkstone watched in horror as a French plane that'd taken off from Croydon plunged into the Channel with such force that it 'smashed to matchwood'. Not every flight even made it that far. On Christmas Eve 1924 a Daimler Airways DH34 setting off from Croydon for Paris struggled to take off in the headwinds and crashed shortly after leaving the ground, killing all eight people on board. Newspapers described in gory detail the 'charred bodies', and the country's first air crash investigation was instigated.

Other incidents were more preposterous than anything else. In 1922, a Daimler DH34 somehow managed to land on top of a *second* Daimler DH34, already on the runway at Croydon, totalling both aircraft. Incredibly, the only injuries sustained by the passengers were a few bruises and a cut lip. In 1924 eight passengers on a Croydon-to-Paris flight looked out of the window mid-flight to see a mechanic on the wing frantically trying to fix a throttle. When he realised mending it was out of the question he remained out there for 45 minutes, holding the throttle open until the plane could land. It sounds like something out of *Looney Tunes*, but this was just another day at the office for those revolutionising air travel.

The upside, if you can call it that, of the numerous air accidents was that planes travelled lower and slower, meaning if you *were* in a crash, the chances of survival were reasonably high. The book *Imperial Airways and the First British Airlines 1919–40* recalls an accident in which two of the propellers went, and the pilot made a forced landing in a field. 'In so doing he collided with a telephone pole and a tree stump. Everyone walked unhurt from the aircraft, astonished to observe how severely damaged it was.'

In August 1931 a Farman Goliath, attempting to land in a typical Croydon fog, smashed into a fence and very nearly collided with a tram. The pilot leaped out of the plane, waved down the oncoming tram, then shook hands with the driver. The aircraft had also very nearly gone through the front room of a house belonging to a Mr and Mrs Manning, and it wasn't long before the pilot had charmed the latter into passing him a boiled egg through the window. 'After this,' reported the *Aeroplane*, 'the pilot, amidst loud cries of "*Vive l'Aviation Civile*," taxied to the tarmac.' It was one of three air crashes inside a day at Croydon.

Often, though, Croydon's persistent fog, compounded by its location at the foot of the North Downs, proved fatal. The ninth of December 1936 witnessed the biggest disaster ever to happen at Croydon, when a KLM airliner attempting to take off in the dreaded Croydon fog crash-landed into a house soon after, killing all 14 passengers and crew. Among them was a former prime minister of Sweden, and Juan de la Cierva, inventor of the autogyro. Herein lies another ghost story. Some time later on a foggy day, a pilot at Croydon Airport felt a tap on his shoulder; he turned around to find the dead Dutch captain of the KLM airliner standing behind him.: 'Don't take off in this fog,' the apparition urged.

Croydonians living close to the aerodrome were understandably anxious. Chimney stacks were getting damaged. Children were frightened. Jokes were cracked about locals sitting on their roofs with a notebook, spotting planes as they swept by as though they were trains. 'We cannot sleep,' complained one aggrieved tenant: 'the vibration shakes every building. The war period was not as bad as this.' And at least during the war it was only bombs falling

CROYDONOPOLIS

on their roofs, not *entire planes*. One gusty February morning in 1927, an RAF plane crashed into a garden in Norbury. 'I just happened to glance out of a window,' said a witness, Mrs Garett, 'when I saw the aeroplane suddenly appear over the roof of a house in Beatrice Avenue, and then plunge down into the garden. It was all over in a second or two.' Milkman William Harris was on his round and rushed over to cut the pilot free from his straps, but he died before he got to hospital.

The top brass were insistent that flying was safe. *Flight* magazine called out the perceived snowflakery of aviation worrywarts; under a section in its June 1922 publication, titled 'Some Absurd Complaints and Suggestions', was the proposal that all planes should carry two pilots as back-up in case one fell ill. This, *Flight* felt sure, would never catch on. Other new safety features were constantly tried out. The *Croydon Times* sensationally reported in 1925 on tests for an 'aeroplane that flies itself', claiming the pilot had left the cockpit and settled into the cabin for half an hour with a good book. Another too-good-to-be-true headline reported on 'crash-proof planes'. The *Croydon Times* must've either been employing some overly optimistic writers, or tinkering with an early form of clickbait.

One safety feature that really did work was conceived in 1923 by Croydon's senior radio officer Fred Mockford. He'd been requested to come up with a standard international distress call. 'SOS' already existed by then, but there were difficulties in distinguishing an 'S' over the radio. Instead, Mockford came up with 'Mayday!', a clear, punchy phrase which sounded to the French ear as '*M'aidez!*' – help me. It was a start. But as aviation's popularity soared, and more and more people wanted to fly from Croydon, its entire aerodrome would have to modernise.

9

CENTRE OF THE UNIVERSE

You could always tell a Croydon native, wrote one journalist in 1933, because they didn't flinch when a plane flew low overhead: 'Not even the largest monsters of the sky will attract the attention of the blasé Croydonian.'

Yet despite Croydon's new status as the aviation hub for the wealthy and the brave, Britain in the 1920s was lagging behind its competitors. Croydon Aerodrome was deficient in technology, aircraft capacity and basic passenger comforts. As an island nation, Britain had wanted to continue protecting its borders from invaders by sea, so had shied away from what it saw as aiding and abetting the technological advances of would-be invaders. Flying was also an expensive business. 'Civil aviation must fly by itself,' Winston Churchill had said. 'The Government cannot possibly hold it up in the air.' But now, as the future of air travel seemed assured, and heavily subsidised national airlines that would become Air France, Qantas and KLM flexed their tail fins on Croydon's grassy runway, the Empire was caught with its pin-striped trousers down. Two things would change all that.

Until 1923 Croydon-based airlines had essentially been 'air taxis': pilots like Alan Cobham and Mildred Mary Bruce hired themselves out to anyone who required their services. But in 1924, Imperial Airways, Britain's first state-subsidised airline, was founded. (It was initially going to be called the British Aircraft Transport Service, until someone pointed out that spelled 'BATS'.) Finally Britain had a proper airline to compete with the rest. To prove it, a race was set up between an Imperial Airways Argosy airliner, the *City of Glasgow*, and the *Flying Scotsman* steam train heading from London to Edinburgh. As it transpired, the *City of Glasgow* only pipped the train by 15 minutes.* Still, there was little doubt: flying would become the superior mode of transport.

Britain's second flagship statement that it was taking air travel seriously came on 30 January 1928, when a new purpose-built airport complex opened on Croydon's Purley Way. 'The suburbs shall shake at the sound of the cry of thy pilots,' one newspaper had captioned a picture of the flashy buildings, quoting the Book of Ezekiel. Here was the biggest airport in the world, and the envy of all other nations.

Walking into the departure hall of the refulgent white neoclassical terminal building, you were greeted by an octagonal clock showing details of flights and departures. A board on the wall detailed what the weather was doing in different corners of the globe (this was soon scrapped when it was realised passengers didn't necessarily want to know they were in for severe turbulence *en route*). It had the first airport shop: chunky teak desks selling cigarettes, sweets and French newspapers. Subsequent airport terminals would opt for a sleeker, modernist aesthetic, but Croydon's late Twenties airport, along with the aerodrome that had preceded it, gave the world a blueprint for everyone else to follow.

The *Daily Mail* painted a vivid picture of the passengers in 1929:

> There was an old gentleman in spectacles, spats, an ulster, and a muffler, smoking meditatively over a literary review ... his grey-haired wife beside

---

* To be fair, the plane had to start from 12 miles south of King's Cross at Croydon, then follow the coast up to Edinburgh rather than going as the crow flies.

The new, purpose-built Croydon Airport complex, opened in 1928 complete with departure boards, weather updates and shops selling foreign newspapers, set the precedent for all international airports to come. *Alamy*

him was making entries in a pocket diary ... near them, a honeymoon couple, evidently Americans, were assuring each other in semi-whispers of a truth obvious to all the world. The middle-aged women in leather coats discussed the relative merits of Italian hotels. Three university students in plus fours argued on Socialism.

Croydon's new airport also had the first purpose-built air traffic control tower, a satisfyingly sculpted hunk of architecture that beamed out across the runways like some dishy chisel-jawed captain. It was kitted out with the latest Marconi direction-finding radio tech, although it was still common for air traffic controllers to step onto the balcony and listen out for incoming planes.

CROYDONOPOLIS

Now more than ever Croydon was somewhere for folk to spend a day out, either to pick out the planes and their pilots, or otherwise swoon over and slander the fashions worn by the wealthy ladies striding across the runway to board their flight. A viewing platform was placed on top of the new airport, and by 1936 was welcoming a staggering 107,059 visitors a year. The Aerodrome Hotel, built next door, installed a viewing platform of its own. With its sprawling lawns and bilingual menus, this was the first purpose-built airport hotel anywhere. Croydonian Donald Gill later remembered spending childhood days watching Handley Pages float down like giant moths at the new airport:

> A man in a peaked cap, white jersey and seaboots used to come out on the balcony and peer at the sky through his binoculars until he could wave the incoming aeroplanes in ... It always used to remind me of the boatman at Wandle Park lake shouting, 'Come in, No. 16'.

Many of those Handley Pages Gill saw would have been Silver Wing flights. A centrepiece of the Croydon/Imperial Airways offering, Silver Wing services departed daily at lunchtime from Croydon to Le Bourget, Paris. Passengers would check in their luggage at Imperial's majestic Art Deco Empire Terminal in Victoria (the building's still there today), before being chauffeured down to Croydon. In those days you could get from the airport entrance to your seat on the plane in a brisk 10 minutes.

The Handley Page HP-42s – sometimes called 'Flying Bananas' due to their shape – bore sonorous names like *Hannibal*, *Heracles* and *Hengist*. Their interiors were based on luxurious railcars, and they sported miniature ensigns poking out of the roof above the cockpit. Graham Coster's *The Flying Boat That Fell to Earth* describes how they 'jounced across Croydon Aerodrome's grass airstrip so enthusiastically that seatbelts needed to be fastened to enable passengers to retain contact with their seats.' Freeman Wills Crofts' 1934 murder mystery *The 12.30 from Croydon* chronicles the thrill of a Silver Wing flight from the viewpoint of a young girl visiting the airport for the first time: 'There was a feeling as if an enormous hand had grasped the machine and was pulling it forward,' writes Crofts as the Handley Page HP-42 takes off – and soon after delivers the back-handed compliment: 'Two or three hundred yards below them Croydon seemed a far prettier place than it had looked on the way from Town.'

As you levelled out above the clouds, white-jacketed stewards, who'd usually honed their craft on trains or ocean liners, offered a choice of four champagnes and 10 cocktails, including the signature 'Silver Wing' (the recipe, sadly, lost to time), and a six-course meal served on blue and white china, with metal cutlery, real glasses and damask tablecloths. While the food had been picked up from Surrey Street market that morning, pastries were plucked from the boulangeries of Paris. Importantly, after a couple of

A Handley Page HP-42 at Croydon Airport: these luxurious (if lumbering) craft were totems of the golden age of air travel with their Silver Wing flights between Croydon and Paris. *Alamy*

glasses of fizz, the latest addition of an onboard toilet was at your disposal. Silver Wing flights were the epitome of luxury (though the French tried to go one better by starting a 'Golden Ray' service), and there was even a 1930 play called *Silver Wings* staged at London's Dominion Theatre, featuring a show-stopping moment in which a plane came through the roof.

There was one snag: each passenger had to be weighed separately before boarding. In the book *Croydon Airport 1928–39* there's an anecdote about a *Daily Mail* journalist of a certain girth having to pay an excess fee, even before his luggage was taken into account. The 25-stone author Douglas Duff had the indignity of being photographed boarding an Imperial Airways flight and dubbed 'the world's heaviest air passenger'.

Le Bourget was just the tip of the iceberg. Imperial Airways routes became progressively more intrepid. By the 1930s, exotic posters were marketing trips from Croydon to New York, Cairo, Baghdad, Calcutta and Karachi (though the full journey would take days, even weeks). Harry Beck, famous for creating the 'circuit board' London Underground map still used today, was commissioned to draw up a similar-style map for Imperial Airways. Instead of Goodge Street, Dollis Hill, Ladbroke Grove and Camden Town, the lines led to Gaza, Delhi, Luxor and Cape Town.

In 1935, Imperial Airways hooked up with Qantas to create the 'Kangaroo Route' between Croydon and Charleville in Australia. This earned its name by stopping at Paris, Athens, Alexandria, Gaza, Baghdad, Basra, Kuwait, Bahrain, Sharjah, Gwadar, Karachi, Jodhpur, Delhi, Cawnpore, Allahabad, Calcutta, Akyab, Rangoon, Bangkok, Alor Star, Singapore, Batavia, Sourabaya, Rambang, Koepang, Darwin, Longreach and Charleville. For good measure, passengers also had to catch a train between Paris and Brindisi.

Cross-Channel excursions, meanwhile, became as casual as catching a train to another part of England. Booze cruises to France had nothing to do with loading up on bargain burgundy in Calais; rather, they were jaunts

The sumptuous interior of an HP-42. Imperial Airways' stewards would often source fresh produce from Croydon's Surrey Street Market. *Mary Evans*

to Paris in which passengers would dress up to the nines, be plied with champagne en route to Le Bourget, get wined and dined in the hostelries of Paris, maybe see the can-can danced in the freshly-rebuilt Moulin Rouge, then be zoomed back to Croydon as the sun came up. Whether you were going long- or short-haul, though, Croydon was firmly at the centre of the world – if not the universe.

Flying didn't come cheap to anyone. Peggy Salaman's record-breaking Puss Moth had been a birthday present from her not-impecunious mum. Even Amy Johnson had to lean on wealthy sponsors. Passengers required a similarly bottomless well of cash. In 1935 a flight from Croydon to Brisbane would set you back almost £15,000 in today's money. If you fancied coming home, it would be the same again.

A secondary market appeared for 'joy riders' – people who craved the thrill of going up in a plane but couldn't afford to be taken too far. Some joy rides were as brisk as being whipped up in the air and circled around the airport (something you could do for as little as five shillings). Catherine Gunn remembers one such joy ride, aged 19: 'Going up was exciting but scary,' she tells me. 'Looking down, all the houses looked like a model village.' For a guinea, the pilot would take you over central London and, if you liked, would spice things up with a few stunts. The concept was wildly popular. In 1921, *Flight* reported that one Captain Muir had had such a busy Sunday of joy riding, he 'did not appear to spend more than five minutes on the ground'. There were 'tea flights' too, another Jimmy Driscoll wheeze marketed in cahoots with Kennards – dishing up tea and cakes in the heavens. 'The day was glorious,' wrote a *Croydon Times* journalist in 1934, invited aboard the inaugural tea flight, 'and cruising along at just over 100 miles an hour the passengers were able to look down upon London and her suburbs shimmering in a haze of heat.'

Fancying a joy ride of his own, the department store magnate Harry Gordon Selfridge chartered a plane from Croydon on New Year's Eve 1934, crammed it with models wearing the latest Selfridges fashions, circled over the capital and, as Big Ben chimed in 1935, had the palooza to end all paloozas.

That same year, the airport welcomed a glamorous new neighbour just across the road: the Purley Way Lido. With its natty Art-Deco-meets-Neoclassical design, it was the first pool anywhere in the UK to be heated and purified entirely by electricity, accommodating a staggering 2,000 bathers in one go. It was also known as the 'Ozone Bathing Pool': advertisements from the time boasted that the purifying gas would delight swimmers with 'that pleasant smell they remember so well from their visits to the coast, with the sea dashing on the rocks and beaches and the wind whipping the spray to a light foam'. Two colour-changing cascade fountains frothed away at each end, while beaches for sunbathing flanked the water's edge, their palm trees swaying in the Surrey breeze. With silver biplanes sweeping low overhead, Purley Way Lido was a scene straight out of an exotic holiday poster, a tonic for those who needed to get away but didn't want to remortgage their home.

The lido became a destination in its own right. International polo teams played here within moments of landing at the aerodrome. Girls in grass skirts did hula dances on floating rafts. Athletes like Helen Orr came to perform two-and-a-half front somersaults off the 10-metre-high diving board. Johnny Johnson, 'England's Greatest Water Comedian', treated crowds to his unique brand of 'aquanonsense'. At times the place was less public lido, more a very wet theatre.

In 1935, the Purley Way Lido opened directly opposite Croydon Airport, visible in the distance. The first pool in the UK to be heated and purified entirely by electricity, it could hold up to 2,000 bathers. *Topfoto*

Croydonians still affectionately (and not-so-affectionately) remember the miasma of chlorine, bunking off school, being dumped by girlfriends, getting thrown into the water, red stomachs from mis-judged dives off the diving board ('Ooooh, that slapping sound! I can still hear it now – *ouch!*'), the woman who sold sweets and chocolates from a pram, and the sunburn everyone got during the heatwave of 1976 ('There were so many bodies in the pool, you could hardly see the water in between them.') Three years after that heatwave, though, the lido was filled in with concrete.

When the new Croydon Airport had opened in 1928, there were 26,000 annual passengers; seven years later, there were over 120,000. Almost half the country's air passengers were using Croydon. 'Whatever else it has done,' wrote one newspaper, 'Croydon Airport has certainly put Croydon on the map of Europe. Many foreigners instinctively think of Croydon when they hear the word England mentioned.' It was quite the change from a few years previous, when passengers had been so thin on the ground, members of one British airline had once had to *dress up as paying customers* in order to deceive an American journalist, who'd otherwise have been the only passenger on the flight.

Despite soaring passenger numbers, Imperial Airways flights only took a handful of people at a time – and that's because they weren't the only cargo. Sending letters and parcels by air had been successfully trialled at the original aerodrome, and by 1930 you could post a letter in London and have it arrive in Paris by air mail just five hours later. In 1934, the Brisbane-born Charles 'Smithy' Kingsford Smith delivered 100,000 letters, including ones in a special royal blue silk bag written by the King, Queen, Prime Minister and Postmaster-General, from Croydon to Brisbane, in what was the first direct air mail from Britain to Australia.

Trade by air was brisk: silks and strawberries came from Paris; tulips (and the occasional consignment of queen bees, each in their own matchbox-sized container) from the Netherlands. In return, Croydon sent everything from furs to Constantinople to cigarettes to Heligoland. It was the beginning of air freight as we know it. Vast shipments of gold were also ferried in and out of Croydon by air. In 1935 Croydon Airport was the scene of a scandalous robbery when over £21,000 (£12 million in today's money) worth of bullion, golden sovereigns and American Eagles was nabbed from the airport safe. The thieves had got into the safe simply by obtaining a copy of the keys, their efforts made even breezier on the night of the theft thanks to one of the guards being sound asleep. The stolen gold was never found, and only one suspect was ever jailed. The safe is still in situ at Airport House and, if you ask nicely on a tour, you can take a look inside.

Crimes at Croydon Airport were often in the news, from cocaine smuggling to arrests on the airfield, including one of a Liverpool youth who'd pinched £20 off his mum, run away from home, spent his last of the funds on a joy ride at Croydon, and was about to walk back to Liverpool when he was caught. 'It is probably that his capture was to his own advantage,' wrote *Flight*, noting that the airport paid for the boy's train fare home.

Croydon also became a muse for one of the greatest crime writers of all. In Agatha Christie's novel *Death in the Clouds*, Hercule Poirot finds himself on a flight coming in to land at Croydon, only for a passenger to die under the usual mysterious circumstances. One of the suspects is a young dentist, and funnily enough, in 1933 – two years before Christie's aerial murder mystery was published – there was a real-life mystery in which the plane *City of Liverpool* went up in flames and crashed shortly after taking off for Croydon from Belgium. A dentist, Albert Voss, had leapt from the burning craft and, though he died, suspicion fell on him for sabotaging the plane in a bid to fake his own death. If that is what happened, then he forgot one key thing: to put on a parachute.

Another real-life episode that was the stuff of a detective novel occurred in 1928, when the well-known Belgian financier Alfred Loewenstein excused himself to the toilet on a Croydon to Brussels flight, and never returned. One of the plane's outside doors was found flapping in the wind, and Loewenstein's body was later washed up near Boulogne. Was it suicide, murder, or an unfortunate accident? No one ever cracked the case – not even Agatha Christie.

10

FALL FROM GRACE

Croydon is dead, her halcyon days are done.
No more her klaxon stirs the summer night.

The death knell for Croydon Airport would come for reasons no-one could have foreseen.

Following the First World War, an idea had been mooted to rename Britain's major airports after battlegrounds. Manchester would've become 'Passchendaele', Howden in Yorkshire 'Cambrai', and Croydon 'the Somme' (you can just imagine the pilot's announcement: *Ladies and gentlemen, please fasten your seatbelts, as we will shortly be landing in the Somme . . .*'). That idea was nixed and, within five years of manufacturing weapons to bring down German planes, Croydon was welcoming them back. At the start of January 1923 a Dornier passenger plane became the first German craft to land voluntarily in Britain since the war. Soon, the first Junkers – muscular, all-metal machines – were coming to Croydon, as a new London-to-Berlin service was in the offing. 'GERMANS COMING,' splashed the *Croydon Times* in another of its sensationalist headlines, adding in much smaller font further down, '(this time on a peaceful mission).' Over the next decade-and-a-half Junkers became a familiar sight over Croydon. Even when Hitler grabbed the reins of power in Germany and swastikas began appearing on the tailfins of these Lufthansa-operated craft, nothing much was thought of it.

But by 1937 there was a growing sense of unease. This was the year in which Hermann Goering, Chief of the Luftwaffe High Command, landed uninvited at Croydon. Goering, a pudgy, preening egomaniac, had decided off his own back that he was going to represent Hitler at the coronation of George VI. The Brits quickly convinced Goering that this wasn't going to be such a great idea, and Goering skulked back to Croydon the next morning to fly home. Germans were still trickling in and out of Croydon up until the outbreak of the Second World War; rumours surfaced that Lufthansa's Junkers Ju 52s had been taking detours over the RAF airfields at Kenley and Biggin Hill to get a few 'souvenir snaps'. It also turned out they didn't need much tweaking to become very effective bombers; in fact, they already had bomb racks on the wings.

On 26 August 1939 the British Ambassador to Germany, Neville Henderson, arrived at Croydon with a message from Hitler that suggested the Führer wasn't backing away from Poland. The swastika-daubed Junkers Ju 52 Henderson had flown in on would become the last German airliner to depart Croydon before Britain went to war. There was another Lufthansa plane still in a hangar at Croydon, but the Germans tasked with flying it back to their homeland found a rivet sticking out of one of the tyres: someone, it appeared, had punctured it. The flustered Germans were forced to abandon the craft, meaning Britain had captured its first Nazi plane of the war. A couple of days later, goes an anecdote in *Croydon Airport and the Battle for Britain*, a couple of British soldiers were ordered to fly the captured plane (still bearing its swastikas) to another RAF airfield, only to suffer engine trouble and make a forced landing in a field. The pilots were duly rounded up by an overly keen soldier and taken into custody. He thought he'd captured some over-eager Nazis.

FALL FROM GRACE

German ambassador von Ribbentrop (left) at Croydon Airport in 1936. German airliners continued to land right up to the outbreak of war. *Mary Evans*

And just like that, Croydon Aerodrome was no more.

Civilian flights were halted. The prize geraniums in the gardens by the aerodrome were torn up. The Aerodrome Hotel and airport hangers became accommodation for soldiers. Handley Pages were out, Hurricanes were in. The RAF was now running the show, and the aerodrome went back to being a military outpost used to defend the nation, and specifically a Fighter Command base that, along with nearby RAF Biggin Hill and Kenley, would play a full part in defeating the endless wave of German attacks in the second half of 1940 during the Battle of Britain. The country was at war, and Kennards printed a rallying ad in the *Croydon Advertiser*: 'Carry

On Croydon,' adding, 'don't worry, it may never happen.' Other voices suggested differently. 'Croydon must beware,' smarmed the Nazi sympathiser Lord Haw-Haw over the airwaves on 26 October 1939. 'She is the second line of defence. We know the aerodrome is camouflaged but we know just what kind of camouflage it is. We shall bomb it and bomb it to a finish...'

He wasn't bluffing. On 15 August 1940, during a performance at the Davis Theatre, the manager took to the stage and announced, 'I have received an air raid warning. Will those who want to leave go out by the nearest exit? The performance will continue.' Meanwhile, passengers on a bus travelling along Purley Way had unenviable front-row seats as three or four planes swooped over the aerodrome and released their bombs. 'They sowed death and reaped ruin ... Houses were smashed to pulp. Explosions thundered in the factory district. Smoke clouds puffed up into the battle-streaked sky,' reported the *Standard Times* in New Bedford, USA, painting a picture in grim poetry. 'Children fled from their play and grown-ups screamed encouragement to busy British pilots and gunners who could not hear above the din.'

The nearby Bourjois perfume factory took a direct hit, filling the air with the sickly scent of 'Soir de Paris'. It was a daytime attack, and there had been no warning siren. Sixty-two civilians died, along with 11 airmen. It was just Croydon's luck that the raid was possibly a mistake on the part of the Nazis. They may well have been angling for nearby Kenley Airfield, but had found Croydon instead. It had echoes of that bungled Zeppelin raid 25 years earlier. The next morning, the book *Croydon and the Second World War* recounts, the factories that'd been camouflaged 'were now gaping black skeletons of twisted, unroofed girders and fallen rubbish from which surged an evil-smelling smoke.'

The raids continued, and others most certainly were meant for Croydon. The longest of these commenced in the early evening of Guy Fawkes Night 1940, and didn't abate until 8.22 a.m. the next day. Hurricanes were scrambled from the airport two to three times a day. Houses and factories around the airport continued to be levelled. Croydon and its airport would never be the same again.

For years, officials had umm-ed and ahh-ed about the suitability of Croydon as a place for an airport. Frederick Handley Page, the man whose airliners had become synonymous with Croydon, wanted a more central London aerodrome. Hyde Park was mooted; so too was the idea of plonking a runway on top of Waterloo or King's Cross railway stations. Such woolly schemes never came to pass, but there was always a sense that Croydon was living on borrowed time. The airport had only really been built there in 1928 because that was the cheaper option at the time. Once Hitler had had his wicked way with Croydon, its appeal was substantially dulled. The airport had been chewed up by bomb damage, and many of its men and women had been lost to enemy attacks.

All the beautiful but now lumberingly anachronistic HP-42s were lost during the war. As planes had got bigger during the inter-war years, grass airstrips had become increasingly problematic in the tropics, where the monsoon season would leave them unusable quagmires, and in 1934, therefore, Imperial Airways' continental flights had ceased flying from Croydon, with all its Empire Air Mail services to be operated from Southampton Water by a brand-new fleet of flying boats. The war had then seen Imperial Airways hastily merged with British Airways Ltd to form the British Overseas Airways Corporation (BOAC), and Silver Wing flights were also a thing of the past.

In any case, the location off Purley Way had never been ideal for an airport, even after new terminal building had opened: there was a rise in the ground on take-off, prevailing winds, boggy terrain, and a number of obstacles to dodge at the south-west corner, which also prevented the extension of the runway. The war had seen planes become a great deal larger and heavier, leading to the adoption of concrete runways across the globe. Croydon's was still grass.

Serious accidents were still occurring too. In 1947, a Dakota C-47A tried to take off in a blizzard, came straight down again and hit another plane on the runway. Twelve passengers perished, including a group of nuns. Sister Hélène du St Sacrement died trying to pull others from the wreckage, her own skirt ablaze. Stories abound of apparitions of a nun on the Roundshaw estate, which now occupies the site of the crash. 'In 1976 a woman on the estate was so distressed at the sight of a nun in her living room,' reported the *Croydon Guardian*, 'that she had to move out of her home . . .'

Other developments were afoot too. On 1 January 1946, the converted Lancaster bomber *Starlight* took off from Heathrow Airport, bound for Buenos Aires. This part of west London, pipped by Croydon at the end of the First World War, was now getting its own back at the end of the Second. In the blink of an eye the capital had a new main hub of international flight: London Airport became official. Croydon hobbled on as London's second airport as best it could, predominantly serving internal and charter flights. But into the 1950s, as housing developments sprung up around its three runways, it was clear this was no place for jet engines. Croydonians understood the magnitude of what they were about to lose, and fought tooth and nail to keep it. 'I feel that Croydon Airport has been neatly planned out of existence by some civil servant's mind,' grimaced local councillor Gerald Southgate. The legendary aviator Alan Cobham was furious, calling publicly for a reprieve. As the 1950s wore on, eleventh-hour attempts were made to save the airport. But when, on 9 June 1959, Gatwick Airport was officially opened by the Queen, everyone knew it was over.

Croydon had been a fulcrum in the growth of continental air travel, and now the explosion of overseas package holidays was on the horizon. But it was time for its last flight. And as any of the volunteers at the Airport museum will tell you with a wry grin, the name of the pilot on this farewell flight from Croydon to Rotterdam was Captain Geoffrey Last. On 30 September 1959, as Last's Morton Air Service Heron swept over the airport, dipping its wings in a valedictory salute, hundreds of tearful Croydonians did the only thing they could: they set light to an effigy of the Minister of Transport and Civil Aviation. The airport officially closed at 10.20 p.m. that night. A poem, 'The Lost Airport', was later penned by Ken Steel, who knew the old airport. It concludes:

> Alas those happy times have passed and gone.
> No more the neon beams its welcome light.
> Croydon is dead, her halcyon days are done.
> No more her klaxon stirs the summer night.
> Croydon is dead, but though her sun has set,
> In airmen's hearts her fame is living yet.

Yet at the very same moment this glimmering era in Croydon's history was being put to bed, another was about to arise. And rise, and rise, and rise.

11

CONCRETE RENAISSANCE

On Wednesday, 15 June 1960, Croydon's life flashed before its eyes. Archbishop Whitgift entertained Elizabeth I with a masque. Roundheads ransacked Croydon Palace. A tipsy William Pitt nearly got shot. Firefighters fought in vain as the parish church burned to the ground all over again. This time, though, the flames were only projections, and the Roundheads were actors. This was Croydon's Great Millenary Pageant, a rambunctious fever dream of the town's thousand-year history, played out by a cast of several hundred in a makeshift area in Lloyd Park. 'I remember the awful weather!' recalls Gillian, then 11, and tasked with playing the part of an English ship fighting the Spanish Armada. 'It rained every evening for the ten days that it was on!'

In the year leading up to Croydon's landmark anniversary, marking 1,000 years since the priest Elfsies had witnessed the signing of a will, and thus recorded the name *Crogdaene* for the first time,[*] enthusiasm was tepid. A young mother, Mrs S. C. Smith, had spent two-and-a-half years

---

[*] An earlier recorded mention of 'Crogedena', dating back to AD 809, was later discovered – meaning Croydon's 1960 millenary date was in fact over 150 years off.

living in a single room with her family, and told a vox-popping reporter: 'Accommodation is more important than squandering money on pageantry.' But the town bigwigs were having none of it. 'This is a MUST,' ran an impassioned advert in the *Croydon Times*. 'The next pageant won't be until 2060!' Sure enough, when the time came, most Croydonians shrugged their shoulders and got into the spirit of the thing, dressing up as fair maidens or sixteenth-century colliers. And now, as the pageant played on a loop night after rainy night, it seemed most of the town had turned out to make historical whoopee. Croydon has long had an insatiable habit of fixing its gaze on the horizon, but for once it was pausing and taking stock.

Except, that is, if you looked close enough at the pageant programme. Anyone who flicked to page 16 would have seen a full-page advert for the Demolition & Construction Co. Ltd, a company that evidently wore its mission statement on its sleeve. In truth, the ad might as well have been a declaration of intent for Croydon itself.

In 1961, the year following the millenary, Alderman Basil Monk, managing director of the Trojan car company, gifted the town with a late birthday present: a viewing platform in the Shirley Hills. High up on this ridge of ancient woodland to the south-east of the town, you could now stand and peer out over the treetops for a panorama of Croydon.

At least, it was *supposed* to be Croydon. Instead, the scene was of a bristling army of cranes poking through great clouds of dust. The likes of the Demolition & Construction Co. were busy rubbing everything out. Strange new oblong forms inched onto the horizon, slab by slab. Croydon 2.0 was on its way.

Croydon had endured a rough war. In 1944, a newspaper reeled off a litany of assaults in the borough: 2,700 high explosive bombs; 2,272 bombs of 500 pounds or under; 44 bombs of over 500 pounds; nine landmines, 94 oil bombs; 202 unexploded bombs. Croydon was hit by more doodlebugs than anywhere

CONCRETE RENAISSANCE

The viewing platform up in the Shirley Hills, gifted to the town in 1961 by Alderman Basil Monk, would soon show Croydonians just how rapidly its skyline was changing. *Graham Coster*

else during the Blitz. Sometimes the missiles simply ran out of fuel en route to London, but double agents had also foiled the Nazis into thinking they were overshooting Central London, with the result that coordinates were tweaked and South London was pummelled instead. Bombing was so intense, some residents formed a wartime version of Neighbourhood Watch, taking it in turns to spend a night or two in the countryside while others on the street made sure incendiary bombs didn't burn their house down while they were away.

Some Croydonians didn't realise how lucky they'd been. One lady was dug out of the rubble by a rescue party and placed on a stretcher. 'Just before we put her in the ambulance', recalled her rescuer, Mr Cook, 'she clapped her hands to her mouth and exclaimed, "Oh, I've forgotten my purse." Despite our protests, she climbed off the stretcher and returned to her ruined home to look for it.' Seven hundred civilians lost their lives to bombs dropped on Croydon, and thousands more were injured.

The mutilation of property was also considerable: 1,400 homes destroyed, another 54,000 damaged. Francis Rossi of Status Quo fame, who remains local to the area, remembered the town as a 'smoggy shithole of bomb sites'. Croydon, like the rest of the country, needed some serious TLC, and it wasn't going to fritter away any time sitting on its hands. By October 1945 Croydon Council had already launched pushy development plans projected 50 years into the future. Widened roads, flyovers, car parks and a host of modern office buildings were all on the docket. 'We insult our intelligence,' wrote the *Croydon Advertiser*, 'if we shrink from the obligations that modern conditions impose. This is a bold scheme because the needs of the town demand boldness.' Such schemes also required someone bold enough to turn them into a reality. Croydon was not disappointed.

Sir James Marshall was such a redoubtable figure it's a wonder he wasn't made from reinforced concrete. A long-standing Croydon councillor and former Mayor of Croydon (not to mention a farmer and a stamp dealer when he had a second or two to spare), this protean Svengali now turned his hand to the role of chairman of the town planning committee. With his austere silver haircut and circular horn-rimmed spectacles he had something of the Demon Headmaster about him, and apparently harboured the same world-dominating ambitions.

Marshall has been described by various peers and pundits as a 'hard-headed autocrat', an 'American-style town boss', 'almost like a mafia boss', 'some evil genius' and – perhaps most tellingly – 'the architect of Croydon'. There's a photograph that shows Marshall looming over a model of the town: he appears omnipotent, all-powerful. That's more or less accurate. 'I believe in Croydon,' he once stated coolly. 'I have always seen it as potentially the most important area of Greater London.' Not since John Whitgift had a lone figure invested so much credence in Croydon – but there was a difference this time. Marshall's vision stretched far beyond a scattering of almshouses and schools. And, unlike any archbishop, he didn't answer to a head of state or god – only himself. In his 1967 book *The Property Boom*, Oliver Marriot wrote that Marshall had once suggested, undemocratically, that 'The best committee is a committee of one.' And so it was that not long after Britain took a $3.3 billion loan from the United States as part of the Marshall Plan, Croydon instigated a 'Marshall Plan' of its own – one with which it vowed to refashion itself in America's hardboiled image.

Strangely enough, Elizabeth II played a role in Croydon's post-war

Sir James Marshall (right, seen here presenting the freedom of the town to RAF legend Lord Tedder) was the redoubtable figure who oversaw the era of 'Croydonisation'. *Alamy*

trajectory, albeit unwittingly. In 1954, the newly minted Queen and her advisers had rejected the town's petition for city status. Croydon was disappointed – humiliated, even. Marshall, certain there was a city inside Croydon begging to get out, was determined to show Elizabeth the error of her ways. In 1956, along with his trusty sidekick, the borough engineer Allan Holt, Marshall pushed the Croydon Corporation Act through Parliament. This was both unique in its scope and pivotal: it gave the local authority the clout to buy up land in the city centre and sell it on, without any checks and balances from the Ministry of Housing. Croydon, essentially, became a law unto itself. It certainly didn't need some wet-behind-the-ears monarch telling it how to run the show. It now had the untrammelled authority to rebuild itself literally from the ground up.

Standing on the junction of Wellesley Road and George Street, the 11-storey Norfolk House looks altogether unassuming today. Its tenants include Travelodge, and Greggs (out of which desperate shoplifters occasionally skitter, clutching Mexican chicken sandwiches). Aside from a set of glorious zig-zag windows currently half-filled with the Wendy's logo, it's nothing out of the ordinary. But when completed in 1959, this was Croydon's first high-rise – the very beginning of Marshall's power trip.

Norfolk House was swiftly followed by nearby Suffolk House, then the loftier Essex House (headquarters of British Rail). Together they became known as 'Little East Anglia'. 'Our grandfathers look at them with a certain amount of distaste,' wrote the *Croydon Times* of this trilogy of newbuilds, 'but our children will regard them with as much affection and unquestioned pride as their ancestors did their own suspect edifices.'

For many Croydonians such changes couldn't have come soon enough. Everyone is familiar with those post-war vignettes of tenement housing from another century no longer fit for use and kids scuttling about on bomb sites. Tastes were changing too. Many in Croydon were thoroughly sick of, as the *Evening Standard* put it, the 'dreary hodgepodge of garbled

CONCRETE RENAISSANCE

Norfolk House on Wellesley Road was Croydon's first high-rise building and, although relatively low by today's standards, heralded the beginning of 'Croydon 2.0'. © *The Francis Frith Collection*

Gothic' and 'Victorian dinginess'. When Robert Atkinson's Neoclassical Croydon Technical College building was unveiled in 1955, the *Croydon Advertiser* poured scorn on this 'mausoleum ... adorned with sham columns and colonnades and other frills borrowed from the age of the horse and cart.' There was a thirst for uncomplicated stacks of concrete and glass, and that's what Marshall was offering in spades (and JCBs).

On the other hand, voices of dissent and prophets of doom were amassing. One Alderman Gibson warned that an unregulated glut of office-building would 'leave a legacy of disappointment and depreciation'. Someone else had suggested that Essex House – which shared little else with its beautiful and iconic Art Deco NYC namesake – was 'rather like a lot of green matchboxes'. This comparison was particularly apt, because the construction of Little

CROYDONOPOLIS

East Anglia had ignited a touchpaper that Croydon would find difficult to extinguish...

'Wherever you go in the centre of Croydon,' wrote Audrey Powell in the *Observer* in 1961, 'there is new building taking place: cranes, skeletons of steel, rising office blocks wreathed in mists of scaffolding, hammering on metal, the hum of cement mixers everywhere. The residents go about their affairs accepting the fact that their town is changing before their eyes.' It was going on before their ears too: the noise was practically deafening, although the *Croydon Advertiser* spun it into poetry: 'a wild orchestral accompaniment of power-charged drills and the coughing snarl of numerous machines ... thrusting Croydon into the twenty-first century.' Stevenage, designated as the first of Britain's New Towns in 1946, had enjoyed much fanfare, and James Marshall reasoned that a wipe-clean-and-start-again approach was what Croydon needed too. As roads were torn up and longstanding landmarks vanished overnight, locals' initial reaction was shock. One resident recalled to *Little Manhattan*, a local project which captures the memories of those who lived through Croydon's regeneration, 'The massive hole in the road was quite devastating, actually, to a child.' Yet as redbrick Victoriana was reduced to a melancholy paprika debris, the altitudinous complexes springing up in its place were one heck of a thrill. 'We were young, it was the dawn of the Sixties, everything seemed to be changing,' recalls John, who lived in Croydon at the time. 'I have to admit I thought these new buildings were great.'

And what an array of new buildings they were. AMP House was purpose-built for the Australian Mutual Provident Society and featured (perhaps oddly for Australians in Croydon) anti-sun glass. Taberner House, toweringly symbolic as the new headquarters of Croydon Council, was likened to Milan's voguish Pirelli Building (although in 2005 it was voted by Channel 4 viewers as one of the buildings they'd most like to see turned to rubble, a wish that's since been granted). RAC House was described by Lord Mountbatten on its opening as 'most imaginative and inspiring'.

CONCRETE RENAISSANCE

Impact House sounded less like the name of a building, more a statement of intent. At the gusty topping-out ceremony of the 13-storey Prudential Building in December 1961, the then-Mayor of Croydon, Catherine Gowers Kettle was lucky not to be blown off the summit. Leon House, its interior concrete spine sculpted with the stunning reliefs of William Mitchell, was a cross between a Wall Street high-rise and the rediscovered remains of some ancient Mesopotamian civilisation. (It soon became the headquarters of Brooke Bond Oxo Ltd, but somehow avoided the nickname the 'Oxo Cube'.)

Up until the 1960s Croydon had been the place from which you commuted to the office. Now Croydon *was* the office. A metropoffice. The big fish of industry and insurance were reeled in all too easily with the promise of low rents, phallic buildings that did wonders for the ego and a speedy

Croydon's council occupied the elephantine Taberner House from 1967 until demolition in 2013. *Suburban Press* artist Jamie Reid once depicted it in the clutches of a destructive monster. © The Francis Frith Collection

getaway at clocking-off time.* Planning applications were rubber-stamped on a near-daily basis. Business turnover and property values doubled, tripled, even quadrupled, in the two decades between the end of the war and the mid-1960s. Oliver Marriot, author of *The Property Boom*, called it 'the most sensational phenomenon thrown up by the office boom in SE England.' A new word was coined in its honour: 'Croydonisation'.

Any capitalist would've been envious of Croydon's new roster of residents: Lloyd's Register of Shipping, Norwich Union, the RAC, Tate and Lyle, Prudential, American International, Rank Xerox, British Steel, British Telecom, British Rail. The biggest coup, though, was Nestlé. In 1964, St George's House – otherwise known as the Nestlé Tower – shinned up into the skyline like a 23-storey block of chocolate, its toothsome Swiss name affirming that, while the airport may have shut up shop, Croydon remained every inch the international player. Allan Holt, the borough engineer who'd helped win Croydon's independence, had highlighted the macho flippancy with which the town was now growing upwards by publicly airing his ambition to build something 'twice the height of Norfolk House'. St George's House was this, and then some. An express lift was fitted so employees could be fired to the top floor in 30 seconds flat. From here, they could drink in views – London one way, the North Downs and Surrey Hills the other – along with their morning cup of Nescafé. The building's height was also magnetic to those with suicidal intentions, including a young woman who leaped from a window during the 1970s. In subsequent years the roof of the nearby bar on which she landed was said to be haunted, the till randomly ringing up receipts for £999.

While from the outside St George's House was a soaring statement that Croydon (literally) meant business, inside it was lucent with marble walls and modernist chandeliers, while its balconies swooshed with delicious curves.

---

* A survey in 1962 showed that 78 per cent of employees at RAC House preferred working in Croydon to London.

St George's House, aka the Nestlé Tower, was a symbol of Croydon's immense commercial success in the 1960s. The borough engineer, it is said, wanted something 'twice the height of Norfolk House'. © *The Francis Frith Collection*

Employee perks included a medical centre and a staff shop selling heavily discounted products, including unlabelled tins ('Sometimes we opened a tin for custard and got tomato soup,' remembers Anne. 'Not so nice on fruit.') Fresh flowers arrived at St George's House once a week and were arranged by women from the typing pool. A chef was poached from a West End hotel, cooking up menus of kidney soup, navarin of lamb jardinière with chateau potatoes and buttered cut beans. After lunch you could recline in the coffee lounge with a cigarette. It was like *Mad Men*, with the old-fashioneds switched out for a nice brew.

The new-look Croydon was soon earning comparisons with New York City. 'The Skyline of Croydon gets more American every month,' wrote the *Croydon Advertiser* in 1965, as 'Little East Anglia' grew into the more imposing 'Little Manhattan'. James Marshall patently had a man crush on Robert Moses, the now notorious 'Power Broker' who'd transformed NYC with major highways, tunnels and bridges, and been called both 'America's greatest town planner' and 'the psychopath who wrecked New York'. Croydon became a regular body double for the Big Apple: in the 1998 movie *Velvet Goldmine* Wellesley Road stands in for a busy Manhattan thoroughfare, the giveaway being that the cars are driving on the left.

At the opening of St George's House, the Mayor of Croydon had joshed that Nestlé could be excused for thinking they *owned* Croydon. Many a true word is spoken in jest; the corporations really were buying the town out. The cupidity of it all was brazen: Marshall and the council were parcelling up Croydon into squares, slapping on upwardly expensive (yet still relatively cheap compared to London) price tags, and raking in the cash. Office rents skyrocketed from 50p a square foot in the late 1950s to £5 a square foot in the late 1960s. Marshall's good fortune seemed boundless: the same year Nestlé's first employees moved in, Labour's Deputy Prime Minister George Brown declared a halt to the construction of new office buildings in Central London. The idea was to decentralise business and disperse office jobs across the towns of south-east England – Woking, Swindon, Slough, Reading. 'Croydon outpaced them all,' writes John Grindrod in *Concretopia*, his study of the rebuilding of Britain after the war, and a kind of foundation myth for someone like him who had been born and raised there. Between 1963 and

1973, 20 per cent of the offices and 30 per cent of the jobs that left Central London wound up in Croydon – insane statistics however you look at them. At this point, James Marshall might as well have stuffed a cigar in his mouth, rubbed his mitts together and growled, 'Come to Papa!' Just as with the archbishops, the Beulah Spa, the airport, and so many other things, Croydon found itself in the right place at the right time. For once the town could take advantage of its murky identity, hovering as some liminal space between London and Surrey. As a motto carved into the front of the Town Hall advised: *Carpe Diem Venit Nox* – 'Seize the Day, for the Night Cometh' …

12

SKY-HIGH SCI-FI

In the midst of Croydonisation, something bigger was brewing. In April 1961 Yuri Gagarin became the first human in space. The following year, JFK gave his 'We choose to go to the Moon' speech, *The Jetsons*, a cartoon about a family living in the year 2062, first aired, and 'Telstar' by the Tornadoes spent five weeks at number one in the UK singles chart. A Ladybird book, *How it Works: The Rocket* – with pictures by Croydonian B. H. Robinson, who illustrated a slew of these children's publications – depicted slender spacecraft soaring up into the ether. From politics to popular culture, the theme was outer space. And at a time when people thought that eating food from tins and quaffing Nestlé granulated tea was the future, it stood to reason that the buildings reflected a similar Space Age philosophy.

On Croydon, two architectural astronauts would leave their footprint. One was Harry Hyams, a young multi-millionaire who gained notoriety when his Centre Point building on New Oxford Street was left standing empty for almost a decade, begging all sorts of questions about the morality of sitting on real estate for profit. Hyams had a coy, boyish outlook, avoiding the lenses of photographers by wearing Mickey Mouse masks. This childlike sensibility was also reflected in the names Hyams gave to his buildings: Telstar House, Space House, Orbit House, Planet House and Astronaut House spoke of adventure and a new age of celestial pilots.

In Croydon, Hyams chose Wellesley Road to speculatively build Lunar House and Apollo House, a brace of ultra-modern office blocks that would soon become the headquarters of the Immigration and Nationality Directorate (IND). The larger Lunar House in particular had touches of NASA about it, thanks to the concrete wings on its roof, like satellites cocked towards the heavens, sniffing out alien lifeforms. Architectural critics like Rosella Scalia have praised Lunar House for being emblematic of the spirit of that age – you can certainly picture it as part of some complex at Cape Canaveral populated by boffins in sleeveless shirts. It's not to everyone's taste, but it's dynamic, different – inspirational. Then again, try telling that to the droves of people who stood in winding queues alongside Lunar House waiting for their fate to be sealed by the bureaucrats inside.

In 2024, West Croydon's London Overground station became part of the newly named Windrush line, in a nod to the Caribbean migrants who came to live in Britain between the late 1940s and the early 1970s. It was a well-intentioned gesture, yet the irony wasn't lost on most people. Croydon's Home Office was a setting of untold misery for scores of immigrants, many of them victims of the *Windrush* scandal.

In 2018, Hubert Howard, who'd moved to the UK from Jamaica in 1960 aged three, visited Lunar House with a journalist to hear from a Home Office official what he already knew: he was not an illegal immigrant. That's what he'd been telling them since 2005, and yet his words had fallen on deaf ears. 'I don't see how an organisation calling itself the Home Office can treat people like this,' Howard told the *Guardian*. 'It has completely ruined my life.'

Three weeks after finally receiving British citizenship, Howard died. Fast-forward to 2024, and it was alleged that Lunar House was the scene of asylum seekers being called in for 'routine appointments', only to be whisked

Lunar House was born in an optimistic era of space travel and Moon landings, although for many summoned here by the Home Office, wrote Les Back, 'The Moon is probably a more hospitable place to visit.' *Alamy*

away in vans as part of the Tory government's much-criticised Rwanda scheme. 'The Moon is probably a more hospitable place to visit,' wrote Les Back in his essay 'So ... fucking Croydon' – not about Rwanda, but Lunar House.

The other man to launch Croydon into the cosmos was Richard Seifert. He was the architect who dragged London kicking and screaming into the twenty-first century with the NatWest Tower in the City, the hulking Tolworth Tower in West London – and Centre Point. As Barnabas Calder writes in *Raw Concrete*, Richard Seifert changed the London skyline more than anyone since Wren, though not necessarily for the better.

Seifert often worked in cahoots with Hyams, but oddly the pair never teamed up in Croydon. Still, he was behind some of the town's most emblematic buildings: his RAC House was demolished in 1989, but Corinthian House on Lansdowne Road continues to stand on its raking concrete legs (a Seifert trademark inspired by Le Corbusier, known as 'pilotis', and still seen in Croydon's newer buildings today), living on as a stylish office space. The structure that would come to epitomise Seifert's tenure in Croydon, though, and become one the country's most memorable structures of this era, wouldn't be topped out till after Neil Armstrong had set foot on the moon.

Pulling into East Croydon on the train for the first time, as Seifert's irregular-looking, icy-white building looms over platform 6, you may find yourself muttering under your breath, 'What's *that?!*' Depending on who hears you, the answer could be: Alpha House, the NLA Tower, the HSL Tower, the 50p Building, the Wedding Cake, the Threepenny Bit or the Bandersnatch Building. Currently, the building's official name is One Croydon.

Opened as the NLA (Noble Lowndes Annuities) Tower in 1970, this was the statement home of the popular pension provider. Its floors stack on top of one another like chunky polygonal coins, which is what earned it its original nickname (and the one you'll still hear from Croydonians of a certain generation): the 'Threepenny Bit'. Squint again, and maybe you'll see some humongous combination lock hoicked on its side that, twiddled the right way, opens up a subterranean version of Croydon called Croydunder.

SKY-HIGH SCI-FI

The NLA Tower was originally received with a mixture of wonderment and horror. As the architectural critics Bridget Cherry and Nikolaus Pevsner cattily put it, you couldn't deny its individuality, but you could deny its architectural merit. But today Croydonians harbour a soft spot for it – at worst, a befuddled respect. The Twentieth Century Society occasionally leads architecture walks around Croydon, inevitably pausing at One Croydon to fawn

One Croydon: icon of the town, building of many nicknames and setting for a fictional spaceport, not to mention one of the most memorable episodes of *Black Mirror*. Graham Coster

143

over the 22-story oddity, while creatives like Gavin Kinch reimagine it in artworks as a towering deck, spinning a giant vinyl on its roof.

The building has inspired entire fictions: Colossive Press, a husband-and-wife publisher of eccentric 'zines based in South London, came up with *Ad Astra Per Croydon*, a hilarious fabrication of a space race between Croydon and neighbouring Bromley between 1965 and 1973. At the centre of the quest to put Croydonaut Norman 'Nails' McAvity into orbit is the spaceport, located, of course, inside One Croydon. 'I was a huge Gerry Anderson fan when I was a kid,' says Colossive Press's Tom Murphy, 'so anything that looks like it might have come out of *Thunderbirds* or *Stingray* is right up my street.' *Black Mirror* creator Charlie Brooker chose One Croydon to star as the headquarters of fictional games company Tuckersoft in his nightmarish interactive Netflix choose-your-own-adventure *Bandersnatch*. The building's time-travelling energy radiated outwards, with St George's Walk, Croydon's long-forsaken shopping arcade, briefly resuscitated with a string of ersatz WH Smiths and Wimpys, as part of an elaborate set for Brooker's show.

Everyone, it seems, wants a piece of One Croydon. There was outrage in 2007 when someone filched a tiny replica of it from a model display of Croydon in the library. (Did the thief love it or hate it? Either way, there was a reason they stole *this*, and not any of the other buildings.) Like an episode of *Black Mirror*, once you've seen it, you cannot scrub Seifert's building from your mind. It adorns T-shirts, prints, postcards, coasters and banners flapping from lamp posts around town. In the half-century since it landed (and landed feels like the right word) it continues to perplex newcomers, comfort denizens and, not least, serve as a highly practical office space.

At the height of its skyscraper saturnalia, Croydon really did seem to be a sci-fi metropolis. The whole of Nestlé's second floor was taken up by IBM's latest computer, and the blue-chip tech company itself later set up offices in Croydon. The town became the biggest user of computers in the UK outside the City of London. Concurrently, plans were afoot for a 'hovertrain' between Croydon and New Addington, the sprawling new estate of mostly social housing constructed at the far end of the Addington Hills, as well as moving pavements for shoppers in the town centre (walking, after all, was *so* 1950s). There were rumours, too, that the newfangled 'self-driving' Victoria line – being bored beneath London that very moment – might extend this far south. Croydon *was the future*. It had come too far to back down now.

The rest of the country was taking note. National newspapers saluted 'Space-Age Croydon ... unequalled through the rest of the country'. 'The comparative happiness of every person in Croydon becomes of national rather than local concern,' wrote Roy Hodson in the *Financial Times* in 1971, 'because the pattern adopted by this Borough in the general manner is being copied more modestly by countless communities wishing to modernise themselves up and down the country.' Meanwhile, as part of a community theatre project, children skipped through the streets of a Croydon housing estate chanting 'Here come the Moon Men!' and 'Take us to your leader!' flanked by all sorts of *Doctor Who*-like creatures – and Pink Floyd were belting out 'Astronomy Domine' at the Fairfield Halls. It was as if Croydon had blasted off and planted itself on another planet altogether.

Croydon's commercial scene had grown distinctly neoteric, too. In the early hours of 7 April 1951 the last remaining Croydon trams – the 16, 18 and 42 – had run for the final time, cheered on by emotional (and in many cases, well-oiled) onlookers. Yet the trams' demise didn't spell the end of Croydon as a shopping mecca – far from it. Post-Second World War, while its department stores continued to do a roaring trade, Croydon began flirting with a world of dynamic shopping experiences, inspired by what was happening in the all-singing, all-dancing United States.

In 1950, the pioneering Sainsbury's on London Road had been gutted of its stained glass and marble and refitted with Perspex and fluorescent lighting. 'Q-less shopping is here!' the store beamed, as it became the first 'self-service' Sainsbury's in the world, moulded in the style already sweeping the USA. 'Self-service' at this time meant that, rather than have a moustachioed man in an apron weigh and package everything up separately, customers now collected a wire basket before weaving their way around the various 'gondola' displays and helped themselves to the produce. 'Every commodity has a price ticket,' explained the *Croydon Advertiser*, 'and when the housewife has completed her tour she checks out at one of five exits, empties the wire basket and pays her bill.' To speed things along, you could even ditch your baby in the pram park, 'over which an assistant always keeps a watchful eye'. Here was a glimpse into a brave new retail world in which the customer was their own servant.

With trams vanquished, cars and buses now started clogging up the streets, and Croydon turned to pedestrianised shopping. In 1964, the same year Birmingham's Bull Ring shopping centre opened, Croydonians were ushered into their new open-top St George's Walk shopping arcade alongside the Nestlé headquarters. Designed by Ronald Ward and Partners, it was furnished with contemporary offerings like a camera shop and – catering for the mania of package holidays which had now taken hold – a Lunn Poly. In the summer you could sit outside the Panino Bar beneath a parasol and have an ice cream; this was America and continental Europe rolled into one. But St George's Walk's foibles would soon come to the fore. 'The wind used to go up against tall buildings and create fearful down-draughts,' recalled one Croydonian for *Little Manhattan*. 'They had to cover it over so that you could walk without being knocked off your feet with the gale.' But it was OK – the Whitgift Centre was about to make its glossy entrance.

When its first shops opened in 1968, the Whitgift Centre revealed itself with a toothy *'Ta-da!'* Its architect, Anthony Minoprio, was a town planner by trade, one who'd already breathed new life into the centres of Chelmsford and Crawley. In Minoprio, Croydon had called in the big guns, and he duly set out to create, in his words, 'a really attractive shopping centre of a kind that has not yet appeared in this country'. He smashed the brief, too. Croydonians ascended a sloped travelator as they were decanted into an effulgent retail Elysium (when the travelator worked, anyway; it was notoriously temperamental). They could nip into Boots – the very first shop to open here – for some Calpol and a *Bridge Over Troubled Water* LP, or buy Space Hoppers and Airfix kits from Zodiac Toys. Fountains splashed playfully in the courtyard, Henry Haig's modernist mosaics lent a touch of the highbrow – and if you were lucky, you'd see a lion cub on a lead being taken for walkies, in what was a particularly edgy marketing stunt for day trips to Longleat. A young Captain Sensible from the Damned (then plain old Ray Burns) bought tins of brown ale from the Sainsbury's before cramming into a photo booth with his pals and taking wacky pictures.

The Whitgift Centre was on the cusp of a revolution that few realised the magnitude of at the time. While Brent Cross is often trumpeted as a landmark shopping centre for London, its escalators didn't start whirring till 1976. The Whitgift, on the other hand, was officially opened by the Duchess of Kent six years earlier. On a walkabout around the shops in burgundy coat and matching wide-brimmed hat, HRH commented, 'Why doesn't every town in the country have something like this?' Good question: the Whitgift had a swoonworthy parade of shops, one people flocked to from near and far, to jettison their hard-earned cash in. It also had the biggest Sainsbury's in the world, boasting 16,600 square feet of goods, a staff of over 200, and late-night openings on a Friday. 'No gimmicks, banners or music to confuse you,' winked Sainsbury's, 'just the widest selection of food in town.' It prompted one Alderman Dunn to proudly proclaim Croydon as 'the greatest commercial centre outside London'.

Of course, the Whitgift Centre wasn't faultless. The critics Ian Nairn and Nikolaus Pevsner concluded that 'most of the architectural details are

CROYDONOPOLIS

The queue to shop at the Whitgift Centre's new Sainsbury's – biggest in the land – on opening day. According to the 2023 film *All of Us Strangers*, the mall was the 'next best thing to Disneyland'. *The Sainsbury Archive*

banal'. It has always been leaky; only partially roofed to begin with, it not only left shoppers vulnerable to rain, but also created another of the town's many man-made wind tunnels that helped earn Croydon the sarcastic nickname 'the Windy City'. One shopper saved the bacon of a frail old woman, managing to grab her just as a particularly violent gust blew her towards a plate glass window. Clearly the Whitgift hadn't learned from St George's Walk's mistakes. Then again, if it did start pouring down, you could always make a dash for the polygonal Forum pub for a restorative bottle of Double Diamond. Local countercultural magazine the *Suburban Press*, on the other hand, railed against this 'new skyscraper tomb town, dedicated to the soulless pursuit of commerce':

On Saturdays the pedestrian shopping precinct 'showpiece' of Croydon, the Whitgift Centre, echoes with the hysterical whine of shopping neurosis . . . You cannot move for shoppers, busily replenishing their property, everyone seems dazed and determined . . . There is always someone to replace you, another coffee and sandwich, another sell.

On the whole, though, the Whitgift was utterly thrilling – another ensign of Croydon's stature. In the acclaimed 2023 film *All of Us Strangers*, in which a lonely TV producer played by Andrew Scott magics himself to 1980s Croydon to spend time with his dead parents, a trip to the Whitgift Centre is described as the 'next best thing to Disneyland'. That's exactly how many people from this part of the country saw it back then.

And yet there was a shrieking irony about Croydon's new-fashioned aspirations. While the town now thrummed with talk of travelators, self-service supermarkets, computers and hovertrains – as John Grindrod puts it in *Concretopia*, 'a town to be approached by jetpack' – Croydon was, in reality, becoming infatuated with a decidedly more middle-of-the-road innovation. The motor car.

13

THE SEVEN HILLS OF CROYDON

The first person to be killed by a car in the UK was from Croydon. Forty-four-year-old Bridget Driscoll was visiting a fete in Crystal Palace on 17 August 1896 when an imported Roger-Benz, taking part in a motor show in the Palace's ground, rounded the corner, zig-zagged about and fatally struck the bewildered pedestrian. While one witness claimed the vehicle had been going at a 'tremendous pace', the driver, Arthur Edsall, said he was only travelling at 4mph, which after all was the speed limit. It probably didn't help that Edsall was driving on the right-hand side of the road, because, as he reasoned, it was a French car. No matter, said the coroner; he hoped this would be the last accident of its kind. This turned out to be an excruciatingly misguided notion: soon it wouldn't just be pedestrians dying in car accidents but motorists too.

And, would you believe it, Croydon was the setting for the first death of a motorist in Britain. As the papers reported in February 1898,

> Mr Henry Lindfield, of Linton House, Brighton, was riding from Brighton to London on Saturday, on a four-wheel motor wagonette, driven

by petroleum with electric ignition, when the car swerved, ran through a wire fence against an iron post, and pinned Mr Lindfield against a tree.

It is grimly fitting that these unenviable firsts are linked to Croydon, because by the 1960s the town had been all but conquered by the car. Yes, Croydon had superb train links (East Croydon was the most-used station in the country outside of London, handling some 700 trains a day), but the future, believed James Marshall, just like his Stateside guru Robert Moses, ran on four wheels. With its time-honoured tradition of widening roads, the first thing Croydon had done in its post-war shake-up, along with the creation of Little East Anglia, was to open out George Street and build a dual carriageway along Wellesley Road and Park Lane, making it reminiscent of Moscow. It was now impossible to imagine that during the eighteenth century this same stretch of road had been a bosky woodland path along which smugglers had lugged casks and chests of tea, liquor and silk under the cover of darkness. This radical overhaul of the road system also featured the Croydon Flyover, which whisked traffic across east and west, plus an underpass to relieve traffic on the new dual carriageway. A local reporter was sent to wade through six-inch-deep gloop and cover the underpass's construction. 'Yet it felt good,' he gushed, 'to see the excavator eating huge mouthfuls of earth and rubble and then spewing them out into the back of a lorry. This was the march of progress.'

While car dealerships formed a ring around the outskirts of town, the local newspapers were choked with ads for Morrises, Standard-Triumphs, Volvos, Citroëns, Fiats, Daimlers, Renaults, Volkswagens, Hillmans... you name it. What's more, Croydon was manufacturing its own cars.

One legacy of the old airport, kick-started by that First World War aeroplane factory that'd barely had time to get off the ground, was the string of factories along Purley Way. Come the 1960s, business here was flourishing, factories churning out everything from Philips televisions to bottles of

During its construction the Croydon Underpass was said to trumpet 'the march of progress', but the strategy of effectively putting cars before people has long since been questioned. © *The Francis Frith Collection*

Tizer. Another successful manufactory was that of Trojan cars. The company was established in 1914 by Croydonian Leslie Haywood, and soon became known for its Trojan Utilities, touted as 'The car for the man who can't afford a car'. When companies like Brooke Bond Tea and the Post Office began using Trojan vans, its stock rose, and by 1962 Trojan had acquired the international rights to the Heinkel bubble car, that cutesy German invention that looked as if it'd been burped out of an alien mothership. In Croydon the Heinkels came pouring off the production line as Trojan 200s. The three-wheeler was nifty, easy to park and gave you bang for your buck. It was also, as an ad in the *Sunday Mirror* pointed out, perfect as a second car: 'Ideal for wife's [sic] shopping'.

CROYDONOPOLIS

Trojan had also acquired the sole importing rights for Lambretta scooters, meaning Croydon was manufacturing a true icon of the mod era: the stylish two-wheeler that spoke of Colin MacInnes's *Absolute Beginners*, early records by the Who, Mods and Rockers duffing each other up with deckchairs on Brighton promenade, and trendy coffee bars.* The scooters sold like hotcakes: 'At one point Lambretta concessionaires were selling more Lambrettas in this country than the Italians were selling in Italy,' David Hambleton from the Trojan Museum Trust tells me. 'As a result of this Trojan's owner, Peter Agg, was awarded the equivalent of an Italian knighthood for his work.'

Perhaps Croydon got a little carried away. A promotional video for an amphibious Lambretta appeared in which a bowler-hatted gent, and a pretty young model riding pillion, wobble their way along the surface of the Thames past Big Ben. 'It's safer than a few seats in Parliament,' reassures the video's voiceover, alluding to the ungainly contraption. Not necessarily the case, as it turned out. On a subsequent demo, the scooter sank in a lake. Mass production was never rolled out.

Croydon's Trojan brand was also to prove an unlikely inspiration for one of the world's best-known record labels. The Jamaican DJ Arthur 'Duke' Reid ferried around his sound system, records and rum in an imported kit van shipped over from Croydon, earning the nickname 'the Trojan', a moniker he later applied to one of his record labels. One of the most acclaimed musicians to appear on the Trojan label was Desmond Dekker, whose song 'Israelites' became the first reggae track to top the British charts. Dekker, as chance would have it, later became a Croydonian when he moved to Thornton Heath.

By 1962 there were six million cars on British roads, with another four

---

* In the late 1950s and early 1960s, Croydon had a number of these coffee bars, places that the powers-that-be associated with juvenile hooliganism. In 1959 the window of the Fiesta coffee bar in South Croydon was smashed twice in two days, although on both accounts it'd been a case of the window being too clean, and people walking straight through it.

During the 1960s Croydon churned out thousands of cute Trojan 200 bubble cars, along with Lambretta scooters. *Alamy*

million predicted by the start of the new decade. Silly amounts of motorists were streaming into Croydon, many for their newly relocated jobs in its jostle of skyscrapers.

In addition to the dual carriageways, flyovers, roundabouts and whatnot, the Croydon Corporation also figured it'd need no fewer than 10 multi-storey car parks in the town centre – space for some 11,000 cars, and then some. You couldn't go anywhere without stumbling on a car park, either already filled with cars or under construction. Terry, a youngster in the Sixties, tells me, 'I remember walking with our late mum down

CROYDONOPOLIS

Sydenham Road, where one of the multi-storey car parks were being built, and all the workmen wolf-whistled mum! She was flattered – she was in her mid-thirties!'

By 1973, Croydon was already onto its sixth multi-storey, sparring with Birmingham to have the most in the country. This wasn't just a point of practicality any more, but a skewed civic pride. In due course Croydon would outdo Birmingham, and the seven car parks built in this period – Wellesley Road, Wandle Road, Dingwall Road, Surrey Street, Fairfield Halls, Lansdowne Road and Drummond Road – would be nicknamed the 'Seven Hills of Croydon'. It was a typically self-deprecating suggestion from Croydonians that their town might be equated with Rome.

Later still, these by now rather shabby, graffiti-ed – and frankly, often urinous – monoliths became an uncanny destination for a series of walking tours, hosted between 2005 and 2015 by former Croydon Council planner Vincent Lacovara. 'I remember finding them fascinating as a child,' Vincent, who grew up in suburban Croydon, tells me,

> when we made family shopping trips to the Whitgift Centre in my dad's Datsun Cherry. As a teenager I learned to drive in Croydon, and would enjoy driving up and down the ramps in my Fiat Cinquecento, and would always park on the top floors, which were generally empty and expansive spaces, surrounded by Croydon's post-war office blocks – which I found exciting.

Vincent's 'Seven Hills of Croydon' tours weren't just about the car parks, of course, but a way to appreciate Croydon from a height, to survey its topography, admire its landmarks and squeeze into panoramic perspective this juggernaut of a town and its herculean intrepidity.

Just as there had been with the surge in building projects, so there was also a backlash against Croydon's fondness for the car. As early as 1958, the

Vicar of Shirley, Canon G. C. Rawlins, had complained that the town had already become one vast car park. 'A situation is fast arising,' he preached in a zippy fable, 'as it has arisen in America, where a man tried unsuccessfully to cross a road. On the other side he saw a friend and asked how he got there. "I was born here," came the reply.' A few short years later, Rawlins' warning had all but come to pass. In kowtowing to the car, Croydon was now an impossible maze to navigate on foot: ramps here, subterranean car parks there, paths leading nowhere. And in what universe was putting shops along a dual carriageway a good idea? The South Norwood architect Owen Luder described the Croydon of 1965 as 'a terrible mess of jumbled, largely

The Fairfield multi-storey to the back of the Fairfield Halls, with Croydon Technical College to the left. Aspiring to build ten car parks would mean the town could say it had more than Birmingham.

unrelated, buildings, and of people, cars and lorries mixed up in a confusion that makes it a madhouse in which to work, shop or live.' Coming from the man who'd designed the unprepossessing Gateshead car park off which Michael Caine had slung an 'out-of-shape' baddie in *Get Carter*, this was really saying something.

Cars were only one part of Croydon's problem. While life was sweet if you were on the Nestlé payroll, in the throes of Croydon's concrete renaissance, something had been forgotten: Croydonians. Six million feet of office space was created in Croydon town between 1959 and 1970 (and another one-and-a-half-million elsewhere in the borough), but it was almost all in the name of corporations and cash, not people. This was in stark contrast to places like Milton Keynes and Stevenage, which were geared towards quality of life for residents. 'There are murmurings from other wards,' wrote the *Croydon Advertiser*, 'that everyone (and by this they mean the council) is so dazzled by the glamorous things that are happening to the centre that nothing is being done for the rest of the town.'

As houses in central Croydon were demolished en masse to make way for Marshall's rapacious enormo-projects, residents were pushed out to fringe settlements like New Addington, whose isolation had soon seen it dubbed 'Little Siberia'. Writes Les Back:

> An intense residential segregation was enforced along the lines of class through building vast council estates like New Addington on the outskirts. These transformations were also coupled with an intensification of racism as Croydon also became a kind of urban frontier for the defenders' racially exclusive Englishness.

Indeed, during the apogee of Croydonisation, a formative National Front was intensely active here, far-rightists often visiting A. K. Chesterton's Croydon flat to discuss their repulsive ideologies.

It's odd to think that, while this disarray of town (non)planning was going on, the tremendously talented Croydonian architect Jane Drew had flown the coop to India, where she was overseeing the construction of the modernist utopia of Chandigarh.

In another life, perhaps she'd have stayed in her home town and rebuilt that instead.

14

CROLLYWOOD

Here is something you might not know about Croydon. It has one of the world's oldest lawn bowls clubs, founded in 1749. It is also the birthplace of Havelock Ellis, an outspoken proponent of autoeroticism and recreational psychedelic drugs. Just as the town's green credentials are so-often overlooked, so too is its predilection for fun – and its tremendous contribution to popular culture.

When in the middle of the nineteenth century *Era* magazine had enthused that 'Croydon is rapidly becoming a most important suburb of London and it has shown greater powers of development than most,' it wasn't talking about just bricks and mortar. Croydon was developing a vibrant cultural life as well.

With trade in the town centre as brisk as it was, owing to Croydon's transformation into a commercial Canaan, so the theatres came, and magnificent ones at that. One by one, a stream of opulent edifices went up: the Theatre (opened 1849), the National Hall and Grand Theatre of Varieties (1895), the Grand Theatre and Opera House (1896). The last of these was a staggeringly ornate confection, with winged angels teetering on its domed facade. It looked as though it belonged not on Croydon High Street, but Vienna's Operngasse.

Indeed, Croydon revamped and rebranded theatres at such a rate, it was almost impossible to keep track. The Theatre opened on Crown Hill in 1849, was reincarnated as the Theatre Royal in 1899, then became the Empire Theatre of Varieties, which in turn became the *New* Theatre Royal. It wasn't done yet: totally rebuilt in 1910, it reopened as the Hippodrome. This particular incarnation was the work of the great Frank Matcham, architect of the London Palladium and the Hackney Empire; Croydon was sourcing the very best man for the job. Nothing was lavish enough for long. There was always room for more crimson velvet seats, more decorative cornices and, in the case of the Theatre Royal, a stage that could be removed to accommodate horses.

Soon the playland of Croydon was giving the West End a run for its money. Sarah Bernhardt performed at the Grand Theatre and Opera House two years on the trot, 1898 and 1899, in front of spellbound Croydonians. (Bernhardt performed exclusively in French, which didn't seem to deter anyone who couldn't understand a word she was uttering.) Vesta Tilley, Dan Leno, Joseph Grimaldi and Henry Irving – some of the most distinguished names ever to tread the boards – all performed in Croydon too. Big-budget staging also lured in the crowds. In 1869 the New Theatre Royal produced melodramatic mise-en-scènes including 'The Poor of the Streets of London' and 'The Great Fire Scene'.

But an even more thrilling kind of theatre was on the horizon. Towards the end of the century the peripatetic showmen were back again, only this time they came bearing gadgets. 'THE RAGE OF LONDON! THE WONDER OF THE AGE!' exclaimed an ad in the *Croydon Times* in 1896, announcing a showing of 'living photographs' at the Horniman Hall. Soon after, the impresario Robert Paul was in town with his 'Theatregraph', screening a slew of short films including *Blackfriars Bridge*, *A Comic Race*,

Queen's Park on Sunday and *Twin's Tea Party*. Said the *Croydon Chronicle*: 'Nothing has been seen like it in Croydon ... The many interesting phases of life, waves moving, people running, trains passing, are all portrayed to the life ...' A new kind of theatre was conquering the world.

The Station Picture Hall in West Croydon became Croydon's first permanent cinema in 1908, screening pictures for a penny.[*] There was clearly an appetite for it because by 1915, according to *The Cinemas of Croydon*, there were *21 cinemas* across the borough of Croydon, each playing three shows a day. At the Rifle Range on George Street business was so brisk it was standing room only.

The first Hollywood movie, *In Old California*, was released in 1910, but Croydon already had its own local film industry years before that. In particular there was the Clarendon Film Company, which started up in 1904, the Rosie Film Company, established in 1906 by cameraman Joe Rosenthal, and Cricks and Martin, makers of *The Pirates of 1920*, and founded in 1908. The latter had a comedy actor on their books called Fred Evans, born the same year as Charlie Chaplin. The two were childhood friends, and in the guise of ham-fisted clown Charley Smiler (later 'Pimple'), Evans was hailed as 'second only in popularity to Chaplin in Britain at the height of his career'. Cricks and Martin were also known for their behind-the-scenes mini-documentaries, in which they showed everything from pottery to Christmas crackers being manufactured. A particularly fascinating film from 1906 was shot at the Peek Freans factory in Bermondsey, showing biscuits being baked, boxed up and shipped out. Another Cricks and Martin film, *The Birth of a Big Gun*, sees the manufacture of a huge naval gun with the use of Hadean levels of flaming, molten metal and next to no PPE. It's enough to make any modern-day health and safety officer keel over in horror.

---

[*] The gabled facade of the shop parade it occupied has miraculously been kept intact, as has the nearby onion-bulb turret of the old Prince's Picture House.

In these early films the onward march of science and industry is often palpable. Percy Stow, co-creator of Clarendon Films alongside Henry Vassal Lawley, was a pioneer of trick photography, and consumed by the idea of electricity. His embryonic 1909 sci-fi picture *Electric Transformations* has a typical scraggly-haired crackpot scientist of the silent era (in this case, a send-up of the real-life electric quack Dr Walford Bodie) harness the newfangled power of electricity to duplicate himself at the expense of one of his young maids. The special effects, which see figures melted down, then revived as someone else, would've had jaws on the floor of the auditorium back then. Clarendon's 1912 caper *The Electric Leg* involves a one-legged man (called Mr Hoppit) who is frankly asking for trouble when he ventures into 'Bound's High-Powered Electrical Limbs' emporium in search of a prosthetic limb. The leg, naturally, has a life of its own, and soon a helpless Mr Hoppit finds himself booting the local bobby through a shop window. Could this have been the inspiration for Wallace and Gromit's escapade *The Wrong Trousers*? Probably not, but it's nice to think otherwise.

One of the joys of watching these films now is that they were often filmed on location, allowing you to see the Croydon of well over a century ago. The Rosie Film Company's 1914 film *The Man Who Never Made Good* offers a whistlestop tour from the bustling Croydon High Street to the edge of Wandle Park, where the eponymous ne'er do well is rather callously attacked by a gang of youths. In *The Nervous Curate*, a hilarious Clarendon comedy that still gets a rapturous reception at Croydon's David Lean Cinema today, there's a scene in which a dragged-up man in the guise of a jealous matron cycles head first into a sewer. At this very moment you also see a tram flash by in the background, a thrilling snatch of incidental Croydon history.[*]

Sadly, however, Crollywood, as it was never called (except on a 2018 film

---

[*] Later on in the same film there's a bit so eerily similar to the 'Here's Johnny!' scene from *The Shining* you have to wonder if Stanley Kubrick didn't stumble across *The Nervous Curate* at some point, and subconsciously or otherwise store this vignette in his head for a rainy day.

location tour of the town), petered out after the First World War, never to become a threat to its American counterpart.

The advent of cinema saw theatres hurriedly change tack. Matcham's Hippodrome was retrofitted with a screen, and became the first cinema in this part of the UK outside the West End to show a 'talkie', namely *The Singing Fool*. In 1928 the Davis Theatre opened, becoming the second-largest

The Davis Theatre was a lavish palace of entertainment which hosted the Royal Philharmonic Orchestra, the Bolshoi Ballet, Buddy Holly and (as its swan song) Ella Fitzgerald. © *The Francis Frith Collection*

cinema in the country (just behind Glasgow's Green's Playhouse), with a capacity nudging 4,000.

The Davis, in fact, decided it could do films, theatre, concerts – the lot. In 1946 the newly founded Royal Philharmonic Orchestra played its first ever concert – at the Davis (you can find this commemorated on a plaque hidden around the back of the fugly building that now stands in its place). When the Bolshoi Ballet did a three-night residency here in 1957, some 5,000 people queued around the block – an almighty boo went up when the management informed those left it had sold out. Operas were another speciality. A young Roy Hudd, whose earliest memory was being held up to the window of 5 Neville Road in Croydon as a six-month-old and shown the Crystal Palace burning down, was once turfed out of the Davis for tittering during a performance of *La Bohème* after Rodolfo started singing to a particularly girthy Mimi 'Your Tiny Hand is Frozen'. Hudd's memoirs are a fountain of theatre anecdotes, another recalling a production of *Sweeney Todd* at the Croydon Empire whose dramatic opening saw the eponymous antihero being chased down the aisle by a brood of Bow Street Runners. On the night Hudd was in, someone stuck their foot out, causing a pile-up of flustered bobbies. 'Who did that?!' demanded Todd. Various audience members pointed towards the guilty party. The Demon Barber duly hauled him from his seat and chucked him out of the theatre, before wiping his hands on his apron and announcing, '*Now* we'll get on with the play!' to rapturous, if mildly terrified, applause.

Box office receipts at the Davis were in another league. On one day pre-Second World War a staggering 11,000 moviegoers filed in through the doors. One of the most popular films ever to screen here was Disney's *Snow White and the Seven Dwarfs*, a showing of which was abruptly paused on 30 September 1938 to announce that Prime Minister Neville Chamberlain had brokered a peace deal with Hitler. The news was greeted with great cheers, although a little over five years later a Nazi bomb came crashing through the ceiling of the same auditorium. Though it failed to explode, it still killed six people and injured another 25.

In 1932 Croydon voted overwhelmingly to allow Sunday screenings, until then a sacrilegious no-no. The unlikeliest backer of this was the Bishop of Croydon himself, a self-professed movie fan. He even made his own film outlining his thoughts on the matter, a cinematic work which *The Cinemas of Croydon* calls 'one of the most tedious short films ever earmarked for preservation'. In these highly competitive times, when a cinema might open one day to find a newer, cheaper one setting up across the road the next, all manner of gimmicks were employed. The first workaday Croydoners to have the curious experience of seeing themselves on film were audiences at the London Electric Hall on Broad Green; vignettes of the town were screened

For the director of *Lawrence of Arabia*, David Lean, Croydon was the bigger desert. *mptvimages*.com

in between short films, and those who spotted themselves were eligible for a cash prize.

Not all Croydon's screenings were at ground level. In 1925 Arthur Conan Doyle's time-travelling dinosaur novel *The Lost World* was made into a silent movie, and the following year became the first ever in-flight film, when a Handley Page loaded with heavy projecting equipment and a select audience of 12 took off from Croydon Aerodrome, flying in a big circle above the town while they watched the screen. The pilot thoughtfully steered through dense clouds to darken the cabin.

Croydon's picture palaces did more than entertain some locals; they became a source of escape. David Lean – not exactly someone, as we know, to hold a flame for Croydon – lived in the shadow of his Quaker father, who believed that things like going to the cinema were a frivolous waste of time. To begin with, Lean experienced the pictures vicariously, listening to the family's cockney charwoman Mrs Egerton describe her own visits to the cinema in detail. Then, one day in 1921, the 13-year-old Lean snuck into Croydon's Scala Cinema with a friend, stumping up sixpence for his first ever picture, *The Hound of the Baskervilles*, starring Eille Norwood. Kevin Brownlow's biography describes how the teen Lean was 'delighted by the huge photographs which moved, and thrilled by the silhouettes of the hound – its body alive with magical phosphorescence.' There and then, Lean knew he wanted to be a part of this screen sorcery. He went on to direct some of *the* classic British movies, including *Great Expectations*, *Brief Encounter* and *Lawrence of Arabia*. All three feature boys or young men who yearn to abandon their humdrum life in search of romantic adventure.

Peggy Ashcroft – immortalised in steel alongside Ronnie Corbett and Samuel Coleridge-Taylor in those sculptures on Charles Street – was a distinguished Croydonian thespian, nominated for BAFTAS, Emmys and Goldens Globes in films and TV dramas including *The Nun's Story* and *The*

Jewel in the Crown. Ashcroft finally scooped a major gong in 1984 when working with David Lean. She'd initially been hesitant about taking on the role of Mrs Moore in what would be Lean's final picture, *A Passage to India*, but her fellow Croydonian wooed her over lunch at the Berkeley Hotel, where they shared reminiscences about their home town. Lean's were presumably a litany of grumbles.

As the decades rolled on, Croydon's grand theatres presented the likes of Maurice Chevalier, Liberace, Paul Robeson and Gracie Fields, who were then followed by a new guard of Brylcreemed rock 'n' rollers with twitching hips and suggestive lyrics. In March 1956 the balconies of the Davis shook when Bill Haley & his Comets blasted out 'Rock Around the Clock'. Buddy Holly and the Crickets came to Croydon in 1958 for a blistering, hip-swinging 55-minute set at the Davis, featuring 'That'll Be the Day' and 'Peggy Sue', the *Croydon Advertiser* bemoaning 'a thoroughly bad-mannered section of the audience'. Perhaps the compère, a whippersnapper by the name of Des O'Connor, could have done a better job keeping them in check. On 8 March 1959 none other than Louis Armstrong was at the Davis – just as Croydonisation was sinking its teeth in, and the Croydon Corporation was about to do something rash, even by its own standards.

15

HERE COMES RUBBLE

Two months after Louis Armstrong was in town, Ella Fitzgerald became the last person to play the Davis Theatre. On 10 May 1959 she performed a mesmerising 20-minute set as part of a 'Jazz at the Philharmonic' gig. Talk about going out on a high. The bulldozers struck the theatre so soon after, Fitzgerald's voice might still have been reverberating around the auditorium. The town of palatial shopfronts and opulent theatre houses sown by the coming of the railways was being returned to whence it came.

Hard upon the Davis being replaced with what one observer called, quite fairly, a 'lumpish shape' of an office block, workers were setting about the dome of the Grand Theatre and Opera House with *pickaxes*, no less, bringing the winged angels tumbling to the pavement. The Scala, Palladium, Hippodrome and Eros theatres were totalled too. (One newspaper's claim that such institutions had been 'outdated by the upsurge of television' had surely jumped the gun.) Other institutions like the Scarbrook Road Baths and the ancient Green Dragon coaching inn also bit the dust.

It amounted to not just an architectural purge but a cultural one, too. Indeed, Richard Rogers would later refer to Croydon as a 'theatre of empty spaces'. OK, this brand of officially sanctioned vandalism was hardly unique in the UK at the time. From Portsmouth to Glasgow, a wave of post-war regeneration was taking hold: ornate theatres, department stores, pubs, churches and shambles superseded by a new style of spartan, po-faced architecture.

But Croydon, being Croydon, went further, and did a more comprehensive job denuding itself of its heritage, than almost anywhere else. Even the pro-Croydonisation *Croydon Advertiser* had questions. 'Is Croydon becoming too dull . . . too stick-in-the-mud . . . too SQUARE?' it asked in 1960. 'YES! scream many teenagers, who miss the star celebrity concerts in which the Davis Theatre specialised.' One employee at RAC House complained that 'Night life is either sordid or nil,' while the Liberal parliamentary candidate Mrs B. M. Bashford lamented that 'On Saturday nights especially one ought to see people out and about enjoying themselves. If you go through Croydon on a Saturday night it's like a morgue.' As someone else put it, the heart of Croydon was in danger of losing its beat.

Croydonians of the future were getting short shrift too. In 1962 the council tried to enforce a ruling that schoolchildren bringing sandwiches in for lunch should pay a twopence 'sandwich tax'. 'These are not the sort of people we want running the town,' sighed one overworked teacher now charged with snaffling the kids' pennies. In the meantime, James Marshall was busy knocking down some schools altogether. Another string to Marshall's ludicrous bow of titles was Chairman of the Whitgift Foundation, and at his say-so Croydon High School for Girls had been toppled to make way for Apollo and Lunar House. But Marshall wasn't done there.

The construction of the Whitgift Centre had meant greenlighting the demolition of the picturesque confection of the Trinity School of John Whitgift – brother school, if you like, of the Whitgift School under the umbrella of the Whitgift Foundation. So eager were they to get started on the demolition in the early 1960s that the last day of term found schoolboys looking on from their classroom windows as the diggers gouged into the cricket square.

In a heartbeat the school's Gothic tower, buttresses sweeping out from its sides, and acre upon acre of verdant playing fields were gone. 'Croydon had either got to deteriorate or go forward,' said Allan Holt grimly. Indeed, Holt even claimed, without irony, that *God himself* had guided him in his vision to revamp Croydon. And yet there was something distinctly sacrilegious about what Croydon was doing to itself. As far back as 1923, in any early bid to widen George Street, the Croydon Corporation had attempted to demolish the Whitgift almshouses, only to be foiled by the House of Lords. You might say Croydon had a self-harming problem.

Not everyone was going to take it lying down. Hundred-thousand-strong petitions were signed. Protests kicked off. Impassioned torchlight parades marched through town. One student at Croydon School of Art

The Tree Walk at the beautiful Trinity School of John Whitgift in the heart of Croydon, the entire site levelled to make way for the Whitgift Centre. © *The Francis Frith Collection*

decided he'd use his newfound skills to hurl spanners into the works of Croydon's commercial overlords. Jamie Reid had enrolled at Croydon School of Art in 1964, and by 1970 he'd co-established the *Suburban Press* (the same publication that had raised an eyebrow at the Whitgift Centre's intentions). 'LO! A MONSTER IS BORN' roared one front cover, an image of the new flyover sinuously piping its way among the freshly planted skyscrapers. Another of Reid's *Suburban Press* works ran the same slogan, but pictured Taberner House, Croydon Corporation's high-rise office gift to itself, being swaddled by a massive, brutish figure – Croydon's answer to King Kong. This 'monster', Reid was telling anyone who'd listen, had got a hold of Croydon Council.

But the most remarkable act of dissent against Croydonisation came from a deceptively meek elderly woman. For a few years, at the foot of the NLA Tower, there stood the handsome Victorian-era East Bridge House, dwarfed by its new sci-fi neighbour. The jarring pairing verged on the comical, and behind it was a story. The NLA Tower's developer – led, uncannily, by a man called John Croydon – was insistent that East Bridge House be flattened to make way for the high-rise's concrete podium.

John Croydon hadn't reckoned with Miss Kathleen Harding, a solicitor who lived and worked inside East Bridge House, and point-blank refused to budge. Croydon (John, that is) wailed to the newspapers about the 'nightmare scenario' Harding was causing – she was thwarting the entire building project, he said. It wasn't fair. The developer tried sweet-talking Harding with the prospect of a virtually free space in the new building. They even offered to call the building 'Harding House' after the solicitor's late father, who'd been Mayor of Croydon during the war.

Harding dug in her heels, to the point where the developers had to build their way *around* East Bridge House – an astounding victory. The house was eventually totalled in 1973, a number of years before Harding had said she'd vacate. But the strange kink of a road that wraps itself around One Croydon

HERE COMES RUBBLE

Victorian East Bridge House sitting defiantly at the foot of the newly built NLA Tower (now One Croydon), after its occupant, Miss Kathleen Harding, had refused to budge. *Croydon Archives/ Museum of Croydon*

to this day serves as a subtle reminder of the time that one woman stood up to the entire concept of Croydonisation and – for a brief moment – won.

Upended and dazed as many felt at the height of Croydonisation, it's reductive to suggest all locals were aggrieved by the metamorphosis of their town.

With the office blocks came a torrent of new wine bars, steak houses and French restaurants, to host a million and one liquid lunches and company dinners – but also cater for Croydonians out of hours. The sound of popping champagne corks ricocheted through the canyons of tower blocks, and employees and Croydonians alike gently steamed in a perma-fug of booze and cigar smoke. Croydon was still, in many ways, the upmarket alternative to the shabbier ilk of Balham, Brixton, Peckham and Streatham. Many of the jobs created in Croydon went to the locals too: 'Half of Croydon worked at the NLA Tower!' one person recalls on a Facebook group.

Some of these roles were more groundbreaking than others. Speak to people like Fay Ruddock and Millicent Reid at the Windrush Generation Legacy Association, and they'll tell you it was difficult for Black people to get a shop job at some of the more upmarket businesses in town during the 1960s. Yet in 1968 Croydon was also on the right side of employment history. Queen's Hospital nurse Sislin Fay Allen was browsing through the classifieds when she came across something that would change her life. It was an advert for the position of a female police officer, based at Fell Road Police Station in central Croydon. She decided she'd give it a go. Given the colour of her skin, it was a surprise to everyone – including one of Allen's friends, who felt sure a Black woman couldn't serve in the force – when she was actually offered the job.

Allen became the first Black female police constable in the country. The magnitude of this wasn't lost on the press, who hounded her for interviews. 'On the day I joined I nearly broke a leg trying to run away from reporters,' she later recalled. 'I realised then that I was a history maker. But I didn't set out to make history; I just wanted a change of direction.'

Milton Keynes was so meticulously crafted the main street was aligned so it framed the rising sun on Midsummer Day. But not even the most basic of master plans existed for Croydon. 'A townscaper is needed here,' the

Sislin Fay Allen served in Croydon as Britain's first Black female police constable. She claimed she 'just wanted a change of direction'. *Alamy*

architectural writer Ian Nairn said in 1964, and yet apparently no one was listening. In just over a decade, the comely Victorian town had receded into a chrysalis of scaffolding, and emerged as some ungainly Brutalist boom town, teetering under its own towering stature and self-importance. Yet, as the architect Renzo Piano once said, 'If a writer makes a bad book ... people don't read it. But if you make bad architecture, you impose ugliness on a place for a hundred years.'

Many of the buildings simply weren't good enough, quickly becoming jokes in themselves. Someone commented that the General Accident building was 'strangely well named.' And while certain architects like Richard

Seifert took unmitigated pride in their superstructures, there was little communal dialogue between any of them. Office blocks had shot up wherever there was space for them to do so, like weeds in an unloved garden. While Croydon looked impressive from a distance, wrote Cherry and Pevsner – almost like a chunk of inner Johannesburg – the closer you got, the more shambolic it appeared. For every Corinthian House and NLA Tower there was a fistful of blunt, lifeless boxes. Rex Grizell wrote in the *Evening News* of a 'Big, rich Croydon with a cold heart', while Peggy Ashcroft ruefully observed that 'The Croydon one sees now is practically unrecognisable.' One resident of the upwardly expanding town during the 1970s and 80s was *Judge Dredd* co-creator Carlos Ezquerra, and there's little doubt that the sparring towers and swirling motorways of Croydon goaded on his vision of Mega-City One, a dystopian metropolis in which his eponymous hero races around stubbing out vice and criminality.

As with its canal, pneumatic railway, its airport, even, Croydon had been so mad keen to prove a point, to put all of its eggs in one basket, that no one had stopped to think things through. While feverishly reshaping Croydon into a futuristic metropolis, had Marshall and his motley band of architects and developers in fact created one of the least future-proof towns there was?

'I think James Marshall sounds like an absolute nightmare,' John Grindrod tells me.

> I think the megalomania – that desire to completely destroy the centre of a town and change it beyond all recognition just because you were personally ambitious – that's not a good reason. And I think there were loads of good reasons to have done this, and actually loads of good things came out of the rebuilding of Croydon. But I almost think they were a bit of an accident.

It doesn't exactly sweeten your view of James Marshall to learn that, while he was presiding over the early days of Croydonisation, he was himself living in a half-timbered mansion called Whistler's Wood on a 200-acre farm in Woldingham, looking out over the undulating North Downs. Out there his bucolic view wouldn't have been interrupted by any skyscrapers.

But at least Marshall had finally given Croydon the clout it required to be called a city? Surely?

While the Brown ban had played straight into the hands of the Croydon Corporation, another radical piece of legislation had since arrived, with mixed results.

On 1 April 1965 the London Government Act became law, and the boundaries of London were redrawn into 32 boroughs that stood under the banner of 'Greater London'. Middlesex was abolished as a county altogether, while the inner parts of Kent, Essex, Surrey and Hertfordshire were sucked into Greater London. Croydon itself was merged with the Coulsdon and Purley Urban District, and absorbed into Greater London – simply as the London Borough of Croydon.

What did all this mean for Croydon? In one respect it had finally achieved what it had always set out to do: throw off the shackles of sleepy Surrey and fuse with the world's greatest metropolis. Not only was Croydon now part of London, but it was also London's biggest town – the *capital* of London, if you like.

And yet. By becoming part of Greater London, Croydon had to face up to relinquishing any kind of Marshallian autonomy – not to mention that it was probably never going to become a city. This became clear in 1965, when another application was rejected. 'Whatever its past history', pronounced the adjudicating board, Croydon 'is now just part of the London conurbation and almost indistinguishable from many of the other Greater London boroughs'.

The Greater London Council now held sway over Croydon, soon putting a stop to its building free-for-alls. Croydon might have walked and talked like a city but, as the whimsical 1971 song by the Danish band Burnin Red Ivanhoe went, Croydon was merely the southern part of a big city very well known.

Things took a turn for the worse. In January 1970 the Conservative leader Edward Heath had holed himself up with his advisers in the Selsdon Park Hotel in the south of Croydon, to cook up a plan to win over the nation. They walked out of that Croydon hotel with a radical free-market agenda. This was what Labour's Harold Wilson mockingly referred to as 'Selsdon Man', a fictional figure who, he warned, was 'designing a system of society for the ruthless and the pushing, the uncaring'.

However, this Selsdon Man was apparently *exactly* what the public wanted, because that June Wilson was voted out of Number 10, and Heath's government moved in. By the end of 1973, though, it was obvious that Heath's tenure wasn't going to plan. Britain was in the grip of 'stagflation', and that Christmas, beset by strikes, especially by the coal miners and the power workers, the Prime Minister lugubriously announced the coming of the Three-Day Week. In Croydon, construction ground to a halt. Offices were forced to power-down. Energy cuts were next. Employees got stuck in lifts and traffic lights failed, bringing chaos to Croydon's new road system. By the end of 1974 queues of Croydonians snaked down the High Street trying to get their hands on a loaf of bread. It looked more like a bleary-eyed communist town than a throbbing organ of British capitalism.

Soon society seemed to be sniggering at the very notion of office culture and high-rises. In 1975, J. G. Ballard published *High-Rise*, a dystopian fiction about a tower block that descends into bloody chaos. A very different kind of satire was first aired on the BBC that year: in the sitcom *The Good Life* Tom Good jacks in his job designing miniature toys to be given away inside packets of cereal in search of an ethical life away from the nine-to-five grind. Nineteen seventy-six saw the launch of another sitcom, *The Rise and Fall of Reginald Perrin*, again lambasting the corporate world, its eponymous anti-hero commuting from his suburban semi to embark on progressively erratic daydreams to escape his dreary employment at Sunshine Desserts. The show

The Selsdon Park Hotel was the retreat at which Edward Heath formulated his 'Selsdon Man' manifesto. It got him elected Prime Minister, but he lasted less than four years.

might as well have been set at Nestlé's Croydon HQ. Everything Croydon had striven for was now being ridiculed.

Croydon, it was becoming clear, had bitten off more than it could chew. The ring road was never completed. The plan for 10 car parks had choked at seven. Now, as almost two decades of cement dust started to settle, it glanced in the mirror and saw a comic punchline smirking back.

Sir James Marshall died in 1979. Unlike so many of the archbishops', his name is not memorialised by any major Croydon buildings, roads or parks. But, to quote the inscription on the grave of Christopher Wren, if you seek his monument, look around.

16

'I COULD BLOODY PLAY THERE ALL NIGHT!'

In a parallel universe, Samuel Coleridge-Taylor was at the opening of Croydon's Fairfield Halls in 1962. Maybe he picked up the baton and conducted the orchestra in a sweeping rendition of *The Song of Hiawatha*. Maybe he returned the following year to watch the Beatles in their matching brown collarless suits get drowned out by scores of bawling fans as they did their best to perform songs from their debut album. But fate, as it turned out, had other plans.

The son of a doctor from Sierra Leone, whom he never met, and an Englishwoman called Alice Hare Martin, Samuel Coleridge-Taylor was born in Holborn in August 1875. By the age of four he'd moved to Croydon, a place where he spent the rest of his life, whenever he wasn't off conquering some corner of the globe. His first biographer, Henry Francis Downing, tells the fairytale story of how a five-year-old Coleridge-Taylor caught sight of a small violin in the window of an old curiosity shop in London. Walking inside as if in a trance, the boy pulled a penny from his pocket – 30 times less than what the instrument was selling for – and to his amazement, the kindly shopkeeper said, 'If you show me that you are a Paganini – that is, if you play me a real tune – the fiddle is yours for nothing.' In his ascent to the highest echelons of musical greatness, via one of the first ever Black scholarships to the Royal College of Music, Coleridge-Taylor would gain comparisons to many great composers, even earning the nickname the 'African Mahler', despite the fact that Coleridge-Taylor never visited Africa once.

From young adulthood it was obvious Coleridge-Taylor was a genius, flying into frenzies of composition in his 'music shed' at the bottom of the garden in his early home on St Leonard's Road. *Ballade in A Minor* became his first smash hit after being performed at Gloucester's Three Choirs Festival at the recommendation of Edward Elgar (the *Enigma Variations* composer championed Coleridge-Taylor as 'far and away the cleverest fellow going amongst the younger men'). His next piece, *Hiawatha's Wedding Feast*, went stratospheric and, alongside two sequels, *The Death of Minnehaha* and *Hiawatha's Departure*, became *The Song of Hiawatha*, which seduced a packed-out Royal Albert Hall in 1900. 'Mr Taylor's work deserves very high praise,' gushed the *Globe*. 'His flow of melody is unfailing, and the brilliancy of his orchestration and the fertility of his imagination are quite astonishing.' Coleridge-Taylor's success as a twenty-something was remarkable enough; that he was a Black man at a time when slavery had only just been abolished, and the British Empire was now encroaching on a third of the African population, nearly inconceivable. Within six years, *Hiawatha's Wedding Feast* alone had been performed some 200 times in England.

America called, and the Croydon composer became the first musician of the twentieth century to crack it. He toured *The Song of Hiawatha* across the States in 1904, 1906 and 1910, stirring up a perfervid fanbase as he went. In Washington DC the Samuel Coleridge-Taylor Society, a 200-strong African-American choir, was formed. Schools were named after him. Teddy Roosevelt invited the young composer to the White House. African Americans rubbed their eyes as the prodigy waved his baton in front of

'I COULD BLOODY PLAY THERE ALL NIGHT!'

Samuel Coleridge-Taylor: composer, conductor, campaigner, Croydonian. *Alamy*

white orchestras. 'His fame', says Jeffrey Green, another biographer, 'blazed like the tropical sun'. Coleridge-Taylor had something else: sex appeal. One cartoon showed the bonny tunesmith hot-footing it from a bevy of fashionably dressed women. 'A distinguished composer escapes from the autograph fiends,' ran the caption. Make no bones about it, Coleridge-Taylor was a rock star, and as such didn't baulk at giving the Establishment the middle finger. At a time when his mentor Elgar romanticised the British Empire to the point where you could sniff the gin and tonic wafting off the score sheet, Coleridge-Taylor became a delegate at the inaugural Pan-African Conference in London, calling for equality regardless of race. Titles like *African Suite Op. 35*, *African Romances Op. 17* (set to the poetry of Paul Laurence Dunbar) and *Symphonic Variations on an African Air Op. 63* also nodded at his heritage. But like many of the great rock stars, Coleridge-Taylor burned a little too brightly.

In the cold, wet summer of 1912, on his way to Crystal Palace's Chinese exhibition, Coleridge-Taylor collapsed on the platform at West Croydon Station. He managed to struggle back to his Thornton Heath home, but died there a few days later, two weeks after his thirty-seventh birthday. Millions went into mourning. 'The world of English music', choked the *Pall Mall Gazette*, 'is poorer today for the sudden, and all too early death of Samuel Coleridge-Taylor.' He'd died of pneumonia, but this had been exacerbated by overwork; the composer had been forced to sell the rights to *The Song of Hiawatha* for a paltry £15.75 (still only about £1,600 in today's money) and, though he managed to secure royalties for subsequent works, was never wealthy.[*]

Over the decades, Coleridge-Taylor was all but forgotten about. The Fairfield Halls put on the odd retrospective, but it was only when Chi-chi Nwanoku founded the Chineke! Foundation, an orchestral organisation for Black and ethnically diverse musicians, in 2015 that the wonders of Coleridge-Taylor were revived to mainstream audiences. 'Remarkably, none of the sixty-two professional musicians in the inaugural orchestra had been taught about, or performed, Coleridge-Taylor's works before,' Chi-chi tells me.

And so, as the Fairfield Halls reopened in 2019 following extensive renovations, both Chineke! and its Junior Orchestra played in the town where

---

[*] This injustice later prompted the formation of the Performing Rights Society, which saw composers paid for the performance of their works as well as recording rights, although this was all too late for Coleridge-Taylor himself. 'The impact his life and death had on music is huge,' Shaniqua Benjamin, Croydon's Poet Laureate, tells me: 'particularly in regards to musicians being paid fairly and getting the royalties they deserve.'

'I COULD BLOODY PLAY THERE ALL NIGHT!'

Coleridge-Taylor had spent much of his life. 'I made a deliberate choice to open our first concert with his stunning *Ballade in A Minor*, marking a commitment to reintroduce and celebrate his compositions,' says Chi-chi. 'It felt as though his spirit lingered in the air, surrounding the orchestra with inspiration. The thought that, had he lived a longer life, he might have witnessed the early days of Fairfield Halls himself adds a layer of poignancy to the experience.'

The Croydon Corporation got lots of things very wrong with its reinvention of the town. One thing it got very right was the Fairfield Halls. When you first clap eyes on this modernist arts centre, you may experience a sensation of *déjà vu*. It is, after all, a dead ringer for the South Bank's Royal Festival Hall.[*] While the architects were different (the Fairfield Halls was the work of Robert Atkinson and Partners, the practice behind Croydon B Power Station – as well as the Neoclassical technical college deplored by the *Croydon Advertiser*), the two concert halls shared the same acoustic designer, Hope Bagenal.

The Fairfield Halls took its name from the 'Fair Field' that once occupied this site and which, centuries earlier, hosted the yearly Walnut Fair, with its legions of roving circuses, theatre companies, oyster and gingerbread peddlers, performing menageries and toy sellers, that drew in 'cockneys of every station and degree'. But Croydon's new pleasure palace was, of course, cutting-edge, and set a new standard that in some respects is yet to be surpassed.

With the Fairfield Halls being completed over a decade after the Festival Hall, Bagenal had the benefit of hindsight. The Festival Hall had been roundly mauled by music critics for its 'dry' acoustics (which weren't significantly improved until a major refurbishment in 2006–7). Croydon's concert hall, therefore, was endowed with a longer reverberation time and fuller tone. It's a point of pride for Croydonians that the Fairfield has better acoustics

---

[*] There are also striking similarities with Coventry's Belgrade Theatre, designed by Arthur Ling and opened in 1958.

than its South Bank *doppelgänger*. The virtuoso conductor Leopold Stokowski praised it as 'the perfect symphony hall', while a 23-year-old Simon Rattle aired his preference for its 'warm and intense' audiences ('You feel so nude at the Festival Hall,' he confided). 'The sound in that place!' gushed the Who guitarist Pete Townshend after thrashing out their new rock opera *Tommy* there in 1969 amid a hypnotic background of flashing coloured slides. 'Oh, Croydon – I could bloody play there all night!'

No expense was spared. Alongside the Fairfield Halls' near-2,000-capacity concert hall was the 670-seat Ashcroft Theatre, the Arnhem Gallery for art exhibitions, plus a Sun Lounge filled with the Chinese ceramics of wealthy local Raymond Riesco. Other bells and whistles included a £23,000 organ, a roll-up CinemaScope screen that appeared at the flick of a switch, and a bunch of state-of-the-art lifts. 'It all makes the Festival Hall seem rather old-fashioned,' wrote one impressed critic. Another even decided that the Fairfield trumped New York City's recent addition, the Lincoln Center.

Some of the Fairfield's mod cons proved problematic. One of the lifts was the cause of a misfortunate episode, when Rosie the Indian circus elephant was so terrified by its juddering that the show she was due to perform in had to be scrapped. 'She is in her stalls having warm milk and a quarter bottle of brandy to calm her,' said Rosie's handler shortly after the incident.

The Fairfield was opened by the Queen Mother in a gala night starring the world-renowned violinist Yehudi Menuhin. 'The diary of inaugural events reads like a name-dropper's manual,' wrote one journalist, noting that Dave Brubeck, the Rotterdam Philharmonic and Chinese pianist Fou Ts'ong were among the first musicians who'd be breaking in the Fairfield's new concert hall.

The Royal Philharmonic Orchestra had made its debut at the Davis Theatre in 1946, but now this had been reduced to rubble, the Fairfield Halls took up the classical baton. Leonard Bernstein led the London Symphony Orchestra (LSO) as they played Stravinsky's *Rite of Spring* here in 1966, and

'I COULD BLOODY PLAY THERE ALL NIGHT!'

Leopold Stokowski considered the Fairfield's auditorium, seen here shortly after opening, 'the perfect symphony hall'. It went on to hosted many of the greatest musicians of all time, from Leonard Bernstein to the Beatles and Bowie.

shortly afterwards Stravinsky himself came to Croydon to conduct. André Previn was often here with the LSO, performing live BBC broadcasts of *André Previn's Music Night*. Carlo Maria Giulini, Sir Adrian Boult and Pierre Boulez all graced the Fairfield stage, while Benjamin Britten chose it to debut his *Songs and Proverbs of William Blake*. These were, without a doubt, the *crème de la crème* – and thanks to Hope Bagenal and some ambitious programming here they were conducting some of the world's finest orchestras in a concrete box by a dual carriageway in Croydon. Croydon, quite simply, was now in possession of one of the finest concert halls in Europe.

CROYDONOPOLIS

The Fairfield's Ashcroft Theatre, meanwhile, had been opened by its namesake in 1962 with a monologue called 'Local Girl Makes Good'. It ended:

> Theatre is a form of giving,
> Croydon I give thanks to thee
> Having named this noble theatre,
> After proud, unworthy me.

The monologue, as it happened, had been penned by John Betjeman, so the great Poet Laureate wrote a poem about Croydon after all – and an uplifting one at that. While the Ashcroft's opening play, *The Royal Gambit*, was something of a misfire (one member of the *cast* referring to the turgid Tudor drama as 'That awful play no one would want to see!'), the theatre went on to host Richard Harris, Bette Davis, Charlton Heston, Rex Harrison, Diana Rigg – the list goes on and on. Versatility was the Fairfield's modus operandi. You could catch a Harold Pinter double-bill one night and watch Giant Haystacks clotheslining Big Daddy during ITV's *World of Sport* wrestling the next. The venue was like a giant, walk-in TV set, one you could flick from channel to channel, from the highbrow to the downright earthy. This felt particularly apt, given how the acrobats, jugglers, musicians, actors and thimbleriggers of those Walnut Fair days had plied their trade on the same patch of soil.

In 1973 the Ashcroft Theatre hosted television greats Morecambe and Wise. 'I've never worked in an aircraft hangar before,' winked Ernie Wise, before the comic duo, wearing smart grey suits and relishing the vaudeville of it all, rattled off a joyous 'best of', 'Little Ern' quips, 'Bring Me Sunshine' and all.[*] All the more remarkable is that somehow this show is the only full live recording of the pair that survives.

---

[*] Alas, their chum André Previn didn't make an appearance for another outing of Grieg's Piano Concerto.

Fairfield's versatility wasn't for everyone. 'I am simply horrified,' one local groused in 1962, 'to think of our Croydon allowing boxing contests to take place in the beautiful new halls ...' But the Fairfield, of course, was never just meant for the pipe-puffing/pearl-wearing classical cognoscenti. To prove the point, rock 'n' roll and all its depraved relations were also decanted into Croydon's modernist hall. In 1962 Jerry Lee Lewis played the Fairfield, his fans point-blank refusing to leave once he'd finished. *'We want Jerry! We want Jerry!'* they screamed, waving their arms and banging their seats. Eventually Lewis returned to the stage to belt out 'Good Golly Miss Molly', which only served to make them go wilder. 'Youths surged around the piano, and jumped down from the back of the stage,' reported the *Croydon Times*. 'An attendant was struck, and three people were thrown headlong from the stage.' When Bill Haley returned to Croydon in 1964 it was to the Fairfield, and the teens gathering by the stage door grappling for a chance to meet their bequiffed idol were so hysterical, venue staff turned the fire hoses on them.

The Beatles caused meltdowns among Croydon's teens not once but three times in 1963. The Fab Four's first appearance in Croydon was at the ABC on 21 March 1963. This was perhaps the last time the group enjoyed any semblance of normality, because the very next day their debut album, *Please Please Me*, was released, and life was never the same again for anyone, least of all them. By the time of their third appearance in Croydon, which was at the Fairfield Halls on 7 September, fans were hurling jelly babies, cigarettes, dolls and money at the group. 'I must remember to take a tin hat with me,' quipped a *Croydon Times* reviewer, although he conceded that John, Paul, George and Ringo certainly had the talent to match the volley of screeching. Again the killjoys were out in force. 'I could not understand why the girls so wished to indulge in such hysteria,' chuntered one such wet blanket. 'Can public disturbances on such a scale be justified and condoned, or must I accept them as a typical expression of musical appreciation emanating from our new Croydon centre of culture?'

CROYDONOPOLIS

While the snobs clutched at their pearls and chewed on their pipes, demanding in no uncertain terms that these feral 'gigs' were banned outright, the management of the Fairfield Halls did the right thing and ignored them. Just as well, because this open-door policy paved the way for an embarrassment of great artists keen to fill the venue with adoring fans: the Rolling Stones, Fleetwood Mac, Ray Charles ('I can only describe this man as enormous – in size, personality and impact', wrote the *Croydon Advertiser)*, Stevie Wonder, Elton John (who during one of his Fairfield gigs playfully slapped his guitarist about the face and hurled a piano stool across the stage), BB King, Ornette Coleman (he recorded a mazy live album here in 1965), Kenny Rogers, Nick Drake (one of around only 30 live shows the troubled folk singer ever played), Hawkwind, T Rex, Genesis, Mike Oldfield, Pink Floyd, Sandy Denny (who sang 'Who Knows Where the Time Goes?' there less than half a year before her death, aged 31), Queen, Sparks, Sister Sledge, Free, Olivia Newton-John. Shirley Bassey performed the same year as the Beatles, and was so spine-tinglingly, palm-wettingly good, said the *Croydon Times*, 'She could have asked for the freedom of Croydon – and got it.' When Kraftwerk played the Fairfield in 1975 it was as if *Autobahn* had been written in honour of the incessant Park Lane traffic whooshing past the front of the venue. And you can imagine that Samuel Coleridge-Taylor was beaming down in approval at the American Folk & Blues Festivals held at the Fairfield in the 1960s and early 1970s, featuring the ilk of Muddy Waters, Howlin' Wolf, John Lee Hooker, Lonnie Johnson, Sonny Boy Williamson, Sister Rosetta Tharpe and 'Hound Dog' composer, Big Mama Thornton.* Muddy Waters liked Croydon so much he also wound up playing an impromptu set at West Croydon boozer the Star. Even the biggest Croydon-basher of them all, David Bowie, couldn't resist the Fairfield Halls. Ziggy Stardust

---

* There's a wonderful photo of Thornton backstage, practising harmonica with a packet of cigarettes and bottle of Johnny Walker close by. Ask anyone where they think this photo is taken and 'Croydon' would not be their first, or indeed hundredth, guess.

and the Spiders from Mars proved such a draw in 1973, Croydon found itself invaded by 1,000 sullen-looking fans, skulking about the streets after they were turned away from Bowie's two sold-out shows.

There was something amusing about this upstanding civic hall – patronised by royalty and lauded by classical greats – doubling up as a house of mayhem and debauchery, especially given that its auditorium's ushers were recruited from the Corps of Stewards, a volunteer fraternity of retired silver-haired men in grey suits. 'It was certainly a surreal experience having one of them show us to our seats for a Status Quo gig, which also had a charity Christmas card fair in the foyer that evening staffed by nice elderly ladies in plaid skirts,' remembers Croydonian Graham Coster. Perhaps those ladies would have enjoyed the gentle Scottish folk ramblings of Alex Campbell, who appeared at the Fairfield Halls in the mid-1960s and, as the book *Rockin' and Around Croydon* explains, was gabbing away to the audience when his false teeth came flying out. Without so much as blinking, Campbell caught his dentures mid-flight, shoved them back into his face and continued talking. 'Howzat!' cried one wit from the stalls.

Croydon had taken a punt on constructing a whacking great concrete casket on the site of an old railway goods yard at one time earmarked for a greyhound track, and turned it into a treasure chest of world-beating entertainment. As it quickly became clear that the Fairfield Halls was going to struggle to make back its investment, some Croydonians branded the project yet another white elephant. But most saw its innate value. 'The Fairfield' was now a name known nationally. 'The corporate soul of the people of Croydon,' wrote one Croydonian in 1967, 'will not be found in vast new shops and office blocks, but . . . in an appreciation of the first-class professional entertainment which recognises the excellent quality of these halls.'

And while Croydon's grandees had been madly kowtowing to the corporates, something else had been going on. In the dingy pubs and clubs clinging on between the rising newbuilds and tumbling theatres, a music scene had been breeding. Mrs B. M. Bashford might've claimed that the Croydon of the 1960s was 'like a morgue', but you've got to assume she was going to all the wrong places. It just wasn't true. And now, as Croydon waded into the wan light of the late 1970s, and the orchestra of drills and hammers tapered off, another sound could be made out. It was the deafening crackle of distorted amplifiers and the primal shriek of recalcitrant young upstarts.

17

SMASH IT UP

At the Star Hotel in West Croydon on the night of 6 February 1967, a gangly figure in a big fur coat was smuggled awkwardly through the back window – not a gatecrasher, but the headline act. So many people had forked out their ten shillings to watch Jimi Hendrix play, the venue was too tightly packed for him to get in through the door.

The legend-in-waiting and his group were already laying down tracks for their debut studio album *Are You Experienced* when they made this improbable appearance (though not as unlikely as the show they'd played a few days before in Bromley's Chislehurst Caves), shaking the Star to its foundations with the distorted refrains of 'Red House', 'Hey Joe' and 'Stone Free' – in short, the future of rock. Plenty of Croydonians remember seeing the Rolling Stones play the same bijou venue, the only problem being that no such gig ever took place. Nonetheless, Croydon was one of *the* places for hip live music.

While the Fairfield Halls might've been the jewel in its cultural crown, the town had long harboured a glorious disarray of venues, clubs and societies. Folk artists sang ballads about Northern Ireland, saxophonists noodled away in the curling Gitanes smoke of the jazz clubs, Beat poets did bits on the sex life of the pandas in London Zoo. Every street corner ebbed with music and literature, mingled with the aroma of late-night chicken and chips, and while Croydon had previously been dubbed the 'Windy City' for very literal reasons, the sense grew that its musical clout might not be all that far off Chicago's after all.

The ABC Croydon in West Croydon, which had hosted the Beatles, and could actually pack in more punters than the Fairfield, had welcomed the Walker Brothers, the Animals, Cilla Black, Cliff Richard and the Shadows and Gerry and the Pacemakers. It was also the scene of a national scandal when, in January 1965, gyrating Texan crooner PJ Proby split his skin-tight bell-bottom velvet trousers midway through the first song. Reports of where exactly the tear started and ended varied: some said it was just at the knees; others that it stretched from knees to the crotch. It almost didn't matter, because this was, as music blogger Russell Clarke later wrote, 'a rip that would be heard around the world'. While some teens swooned at the sight of whatever bonus parts of Proby were now on show, the crotchety moral ombudsman Mary Whitehouse got wind of the incident, decided it was the most outrageous thing to happen since the 120 Days of Sodom and declared Proby's act 'obscene' (not that she'd actually seen it, of course). The press happily latched onto Whitehouse's histrionics, and Proby was banned from all the UK's major theatres, as well as by the BBC and ITV. His career imploded, and he ended up doing jobs ranging from janitor to muck spreader.

Close to the ABC was the Top Rank, where Curtis Mayfield, Jimmy Ruffin, Al Green, the Crystals and Bob Marley and the Wailers gigged. Later rebranded as Cinatra's (the spelling, you'd assume, to swerve litigation from Ol' Blue Eyes), this was the club where Liverpudlian legend Billy Fury sang live for the last time, during a recording for the Channel 4 show *Unforgettable*, before dying of a heart attack a few days later.

Further south, Purley's Orchid Ballroom had started life in the 1930s as an ice-rink, before transformation into the demure setting for big bands, tea dances and fake palm trees that had been the backdrop against which older generations of Croydonians had gone courting. But into the 1960s and 1970s these had gradually made way for the likes of Percy Sledge, Ben E King, the Ronettes and Patti LaBelle. Otis Redding played the Orchid in 1966, wearing a red silk suit, belting out 'Respect' and 'I've Been Loving You Too Long'. 'His act was pure catharsis,' the journalist Mick Brown later remembered; 'not a word I'd have understood then, but you didn't have to understand it to feel it. It was called soul music for a reason.'

Croydon was seminal for so many musicians of this era. The Yardbirds, who counted among their number a fresh-faced Eric Clapton and Jeff Beck, cut their teeth with a long-running residency at the Star's Crawdaddy Club. Joe Strummer grew up in the leafy Croydon hinterlands of Warlingham. A young singer who went by the name of Mandy Doubt worked at Croydon's famous Bonaparte Records, sifting hungrily through the LPs for inspiration and later switching back to her real name, Kirsty MacColl. Not all aspiring musicians would go on to great things – at least not in the world of music; a young flautist travelled to Malta – his first ever trip abroad – with the Croydon Youth Philharmonic. But ultimately Keir Starmer's fingers proved a little too chunky, and he fell into law and politics instead.

For others, Croydon was altogether redemptive. Mott the Hoople were on the verge of splitting up in 1972, but when they had a particularly good gig at the Fairfield Halls, decided to stick things out a little longer. Their song 'Saturday Gigs' recalls how they'd been, as they put it, slipping down snakes into yesterday's news and ready to quit – and then they went to Croydon. There's a similar story about the Electric Light Orchestra. In *1–2 Cut Your Hair! The Johnny Moped Story*, Simon Williams writes how ELO were on a disastrous tour that was haemorrhaging money. By the time they reached Croydon in April 1973, frontman Jeff Lynne had all but thrown in the towel. 'Well, lads, I think this is it. I think this is the end,' Lynne announced glumly. The band duly went out to play to a dozen people sitting cross-legged on the floor. It was certainly looking like a career-ender. That is, until two blokes appeared out of the blue, and started doing some 'mad rock 'n' roll dancing'. Everyone got to their feet, the dance floor filled up, and ELO wound up playing three encores. It was the best gig they'd ever done. Needless to say, they didn't split up that night. Without Croydon, then, there would never have been 'Mr Blue Sky'.

In the mid-1970s, however, Croydon's long-haired rockers suddenly found themselves being barged off the bill by groups of truculent upstarts. 'Hair was getting shorter and music was becoming faster!' says Peter Fox, who DJed across Croydon venues in the 1970s and whose promotions became a fixture of Croydon's punk scene. These sneery young pretenders had names like the Wasps, the Vibrators, the Stranglers and the Slits, and attitudes to match. They swore. They spat. They were angry with the state of the world. At least, that was their schtick. Stuart Bell was 16 in 1977, and used to slather punk slogans like 'London's Burning ... With Boredom' on his T-shirt using little pots of Airfix paint, before sneaking off to gigs without telling his parents what he was up to. 'Suddenly all these bands that you'd read about in the *NME* were playing opposite the Croydon Underpass!' Stuart tells me. 'It was very, very incongruous actually. And very convenient.'

The throbbing centre of Croydon's punk movement was a comically inappropriate venue: a first-floor ballroom in St George's House called the Greyhound. The Greyhound borrowed its name from the coaching inn on the High Street, but this was a very different kind of hostelry. With its thick carpet, velvet curtains and chandeliers, the ballroom could have been designed for Rotary Club dinners, but from the mid-1970s it was opening its doors every Sunday evening for pub rockers and hippie bands. Punk turned it into a grimy fleshpot bristling with surly teens. Sweat ran down the walls and flob sailed through the air (the better a band was, the more they were gobbed at).

Captain Sensible recalled to *Deep-London Magazine* that fee-dodging interlopers like himself would steal in through the fire escape windows and land in people's laps in the dark, prompting shrieks. When the punk audiences weren't rioting on the stage, they were getting off with each other underneath it. 'You had to go to school the next day having pogoed yourself stupid the night before,' says Stuart Bell. That was, until it was banned. Posters pasted up outside the venue read:

> Pogo dancing or similar is not allowed due to accidents & injuries during recent weeks.

The first night the Greyhound tried to enforce this, the Damned were playing. Within seconds, the whole room was jumping up and down like an army of jackhammers.

Mayhem it might've been, but the Greyhound was one of *the* places on the circuit to get your fix of punk, a tastemaker alongside the likes of the Roxy and the 100 Club in Central London. It played host to the Lurkers, Ian Dury, the Adverts, Magazine, the Buzzcocks, X-Ray Spex, Slaughter and the Dogs, the Boomtown Rats, Adam and the Ants and the Cure. In June 1977 the Jam took to the stage, Paul Weller chopping notes from his Rickenbacker before snapping a string and slinging the guitar through the air. 'These guys are good, real good,' wrote the *Croydon Advertiser*.

Even po-faced New Yorkers the Ramones knew the Greyhound was worth playing. 'It was the best gig I've ever seen in my life,' BBC journalist and erstwhile New Waver Rob Plummer tells me. 'I can remember Joey Ramone coming on and going: "Hey! We're the Ramones! If you've got a loud mouth, bebbe, you'd better *shut it up*! ONE-TWO-THREE-FOUR . . ." And they never stopped after that. It was *relentless*.' They were supported that night by an outfit yet to release their first album called Talking Heads.

In that Greyhound melée you might have seen a jejune Norman Cook, aka

CROYDONOPOLIS

GREYHOUND
PARK LANE CROYDON
★ Surrey's Premier New Wave Venue ★

Sunday 22nd Jan	XTC +SECRET
Sunday 29th Jan	ULTRAVOX +ADAM & THE ANTS
Sunday 5th Feb	TALKING HEADS +DIRE STRAITS
Sunday 12th Feb	SIOUXSIE AND THE BANSHEES
Sunday 19th Feb	SHAM 69
Sunday 26th Feb	ADVERTS +ALT. TV
Sunday 5th Mar	RICH KIDS

● PLUS! Peter Fox Lites and Sounds ●
LICENSED BARS   MEMBERSHIP   7 pm
P.T.O.

Sunday nights in the mid- to late 1970s saw a cataclysm of punk and post-punk bands let rip in the sweaty surrounds of the Greyhound.

Fatboy Slim, at his first ever punk gig, which was Generation X here in 1977, a glazed-eyed Shane McGowan looking like he was having an epileptic fit on the dance floor. McGowan, then going by the name of Shane O'Hooligan, subsequently supported the Damned at the Greyhound with his band the Nips, departing the stage furiously flicking V-signs at the audience.

The venue wasn't just somewhere to discover new music, but for new bands to *be* discovered. 'One week a support band such as Generation X would be playing for five pounds,' Peter Fox tells me. 'Then they would appear on *Top of the Pops* and the following week they would be headlining on five hundred pounds.' Post-punk group the Fall were signed for their first Peel session at the Greyhound; it was to be the first of 23 sessions they recorded for the great BBC Radio 1 DJ. Sixousie and the Banshees, who hailed from neighbouring Bromley, which lacked a comparable music venue, and were therefore often to be seen gigging around Croydon, began their Greyhound show in May 1978 with their frontwoman muttering, 'No, we haven't been signed. This'll put the shits up him, though …', before launching into a grungy 'Helter-Skelter'. The show must've done something to someone, because a few weeks later the band were on the books of Polydor.

Siouxsie's Bromley was typically a more wholesome London suburb but, alongside Croydon and other satellites of the Big Smoke like Woking, left quite a mark. 'The London suburbs were very important in the growth of punk,' Rob Plummer tells me. 'There are a lot of places where people felt that life was a slow, suburban death, and they wanted to do something about it.' In 'So … fucking Croydon' Les Back claims Croydon's ambience directly goaded on a climate of rebellion: 'Croydon's sterile shopping centres and council estates provided the ultimate expression of capitalist modernism. Yet they equally provided the perfect canvas for Situationist slogans, radical gestures and the stylistic refusals of punk.'

Adrian Winchester from the David Lean Cinema sees it differently. He was in a punk band called the Bad Actors, and reckons many of the musicians at that time held a certain reverence for their home town: 'If you're to look at any references to Croydon from Croydon musicians, I think you might find evidence of a quite affectionate response,' he tells me.

> If you come from Croydon, you're quite used to it being the butt of certain jokes and being treated in quite a negative way, and so maybe that makes you a little bit defensive about it. We had a song called 'Croydon Girls', that was not really about Croydon, but it was very complimentary about girls from Croydon!

Inebriated on the fumes of anarchy, Croydonians like Adrian cobbled together punk bands in their droves. The 2022 compilation album *Are They Hostile?* is a cornucopia of local groups from back in the day – the Daleks, the Marines, Slime, the Straps, Case – and many of the songs on the record are long-lost gems.

The most infamous Croydon band of that time was Johnny Moped, led by the eponymous frontman (though 'led' is a strong word, seeing as he was often AWOL, and occasionally had to be kidnapped by his band members so they could get him to recording sessions). The band's career reads less like a rock 'n' roll diary, more like an episode of *The Young Ones*: Hell's Angels threatening to de-limb them, matriarchal girlfriends twice their age shooting disapproving looks from the sidelines; on-stage punch-ups in front of record execs. Johnny Moped did, however, create an unpolished gem of a punk record called *Cycledelic*, distinguished by an improbably sentimental pop ditty called 'Darling, Let's Have Another Baby', later covered in a duet by Billy Bragg and Kirsty MacColl. Inevitably doomed to the pigeonhole of 'pretty bloody niche punk', the Mopeds were, however, an effective nurturing ground for some massive talents, albeit not in the most orthodox manner.

For one thing, those 'mad dancers' at the ELO concert had been Moped members guitarist Xerxes and bassist Ray Burns. Johnny Moped also made the rash decision of firing the band's only female member, a certain Christine Ellen Hynde from Akron, Ohio. Hynde, perhaps you've guessed, preferred to be called Chrissie, and went on to form a band called the Pretenders. As for that ELO-redeeming bassist Ray Burns: he would soon put together one of the greatest punk bands ever.

Burns was cleaning toilets at the Fairfield Halls when he met Christopher Millar; the two hit it off with their love for music, and a shared ambition to switch out bog brushes for rock 'n' roll instruments. With their names changed to Captain Sensible and Rat Scabies (the latter derived from a run-in with a rodent, the former, acute irony), the two joined forces in 1976 with singer David Lett (aka Dave Vanian, a play on 'Transylvanian', owing to a brief career as a gravedigger) and guitarist Brian James, to form the Damned.

Here was a cataclysmic whirlwind of noise and velocity. Vanian, face deathly white and eyes rimmed in kohl, would totter onto the Greyhound stage as though just risen from the grave for the band to hurtle into rat-a-tat numbers like 'Neat Neat Neat' and 'Born to Kill', Scabies' explosive drumming would ricochet around the kit, and Sensible would cavort about the stage wearing anything from cricket whites to ballet tutus to nothing at all.

His trademark red beret and circular sunglasses developed as a form of rock 'n' roll PPE. 'It's not very nice having gob in your eyes,' Sensible

Captain Sensible (left) and Dave Vanian (right) in the Damned's punk heyday. Sensible and drummer Rat Scabies had decided there must be more to life than cleaning toilets in the Fairfield Halls. *Redferns, Getty Images*

reasoned, understandably. At their explosive Greyhound set in January 1978 the Damned had, said the *Croydon Advertiser*, 'all the panache of a comic strip cartoon'. By the time Vanian had leapt into the feral crowd the incongruous chandeliers were festooned with discarded clothes.

All gimmicks aside, the Damned's music was good, real good. 'You are forced either into taking them to your heart or turning them away repulsed,' effused Tony Parsons in the *NME*. 'The choice is yours. I've made mine.' Like Croydon itself, the Damned marched forwards, fast. They were the first British punk band to sign a record deal and, while fellow New Wavers like the Clash and the Buzzcocks would disappoint punk purists by signing with multinational record labels, the Damned went with the anarchic independent Stiff, which would soon go on to launch, among others, Elvis Costello and Ian Dury. (Stiff's first 'package tour', featuring both, would achieve the laudable distinction of getting itself banned from a return visit to the Fairfield Halls after all the mayhem that accompanied it.)

The Damned were then the first punk band to release a single ('New Rose', a bona fide belter which still stands up as one of the most electrifying punk anthems to this day), and the first to release an album, the blistering *Damned Damned Damned*, on the cover of which the foursome appeared plastered in cream pies.[*] They became the first British punk band to tour the United States, once again making Croydon an international export. A later hit, 'Smash It Up', offered a fitting anthem for Croydon if ever there was one. The Damned also became the first punk band to split, then reform again. They've been around on and off ever since. In 2016, they launched their fortieth anniversary tour and, as of 2024, are still smashing it up around the world.

It's surreal enough that this tempest of punk was happening in a plush-carpeted ballroom at the headquarters of a Swiss chocolate multinational – even more so when you think that across the road at the Fairfield

---

[*] The modest first pressing of the album featured – deliberately by mistake? – an image of pub rockers Eddie & the Hot Rods on the back cover. These now sell for serious money.

Halls Matt Monro was crooning 'From Russia with Love', or – God forbid – Rolf Harris was puffing away on a didgeridoo at one of his terrifically popular Christmas shows.

One punk group who never played the Greyhound, however, were the Sex Pistols. They very *nearly* did, until Howard Bossick, the Greyhound's manager, nixed the show at the last second, apparently because he didn't want to be spat on. (You'd've thought this was par for the course running a punk venue in the late 1970s.)

But the Pistols are part of Croydon's hall of fame all the same. In 1968 the band's manager, Malcolm McLaren, had attended Croydon College of Art – where he sketched the skyscrapers in charcoal – conjuring up the New York City he longed to visit. It was at the college (whose alumni also include 'Op-art' legend Bridget Riley and Kinks frontman Ray Davies) he became pals with anti-Croydonisation warrior Jamie Reid, which led to Reid landing the gig to design the sleeve for the Pistols' first – and, as it transpired, only – studio album, *Never Mind the Bollocks, Here's the Sex Pistols*. With its screaming Dayglo pink-and-yellow colours and 'ransom note' typography it remains one of the most recognised (and pastiched) album covers of all time, and it's a product of Croydon.

18

TRUE BRIT

The Greyhound was shuttered amid rumours that the gig-goers had been pogoing so seismically that cracks were appearing in the parade of shops along St George's Walk below. Punk was on its way out, and the Damned, Chrissie Hynde and Kirsty MacColl marched onwards into the echelons of fame, fortune and *TOTP*. Yet again, Croydon had held something white-hot in the palm of its hand, and yet again it had fizzled out.

But the town still had plenty to keep it occupied. Artists like Eurythmics, the Human League and a-ha poured in with their tartan suits and sculpted 'dos. Venues like the Cartoon and the Underground, in a basement venue earlier occupied by the ABC self-service restaurant, continued to dish up a rich menu of live music with performances from the likes of Half Man Half Biscuit, the Fall and the Stone Roses. The Cartoon was so-called because it had drawings by some 20 cartoonists slathered over its walls. In 1984, American blues legend Bo Diddley played to a full house here, who couldn't quite believe their luck. Mungo Jerry were Cartoon regulars, the band's leader Ray Dorset often pulling up outside in his white Rolls-Royce and coming in for a drink.

CROYDONOPOLIS

It's not Rolls-Royces the Cartoon will ultimately be remembered for, however, but motorbikes. The club was the hangout for Croydon's bikers, and in 1985 one disgruntled punter who'd been barred decided she was going to get into the Cartoon anyway – namely by driving her motorbike at full pelt through the front door. 'There was an almighty crash as the bike shattered the front door of the busy Cartoon,' reported the *Daily Express*. 'Tables laden with drinks were sent flying, and a regular seated on his favourite stool lost his pint as he ended up on the floor.'

A liberal scattering of record shops – Bonaparte, Gooses, 101 Records, H. &R. Cloake – gave musos their fix of the latest Peel-endorsed EPs and American cut-outs. Beano's on Middle Street was the largest second-hand music store in Europe.[*]

Reggae had taken off in the punk clubs, with DJs like Peter Fox spinning records in between sets, as well as in clubs like the Georgian in East Croydon, which starred one of London's first Caribbean sound systems. In 1979 a pioneering dub artist by the name of Mad Professor set up a four-track recording studio in his Thornton Heath living room, cooking up Ariwa Sounds, a label that would go on to record with Lee 'Scratch' Perry, Grace Jones and Massive Attack. Like Manchester or Seattle, Croydon is one of those places that has music in its bones. Stroll through Wandle Park and you might hear some fledgling rapper spitting bars in the bandstand. For years, one of the big attractions of the North End was Bernard the Reggae King, a street performer who'd draw circles of people around him as he body-popped to tunes from a ghetto blaster.

The fabric of Croydon seeps into popular music. The folk singer Ralph

---

[*] It later sold 8,000 records in one day, to be used as props for the Richard Curtis film *The Boat That Rocked*.

McTell, who'd grown up in Croydon,* gleaned inspiration from Surrey Street; the old man in the closed-down market kicking up the papers with his worn-out shoes in *Streets of London* is a character McTell had actually seen here. 'The architecture of Croydon definitely influenced our music,' Bob Stanley of the pop group Saint Etienne (who, as it happens, once sold eggs on Surrey Street Market) admitted. 'The concrete and scale of it went with electronic music.' Nadia Rose, whose one-shot video for *Skwod* sees her and an Adidas-clad entourage gambolling down Surrey Street, honed her skills on Croydon's streets, scribbling lyrics on the betting slips in the Coral where she worked while hanging out in the Whitgift Centre. 'Something would always inspire me,' she tells me. 'When I left my house and walked down Whitehorse Road, or I'd go to a Crystal Palace football game, there'd always be a trigger. I don't know, Croydon just has this way about it.' Captain Sensible squiffily recorded his number-one hit 'Happy Talk' in a South Croydon studio having generously refreshed himself in Surrey Street's Dog & Bull beforehand. Meanwhile, his nostalgic ode to the stomping ground of his youth, called simply 'Croydon', looks back fondly on his toilet-cleaning days, crooning that in Los Angeles he'd still be dreaming of Croydon, 'especially', he was at pains to emphasise, in the cold and rain.

But as punk faded away and the 1980s came out to play, one North Croydon neighbourhood was about to leave a lasting legacy on music that

---

* Another of McTell's melancholy hits came from a tragic story that unfolded in Croydon. 'Bentley and Craig' recalls the bungled robbery of the Barlow & Parker confectionery company warehouse on Tamworth Road in 1952 by 19-year-old Derek Bentley and 16-year-old Christopher Craig. The latter fired the shot that killed police officer Sidney Miles, and the infamous phrase that emerged from the subsequent trial was 'Let him have it, Chris!' – the words Bentley uttered to Craig just before he fired the gun. Did Bentley mean let him have it, as in 'Shoot him', or 'Let him have it' as in 'Give him the gun'? The jury concluded it was the former, and the teenager was executed at Wandsworth Prison. 'My mum knew the Bentleys,' McTell recalls. 'I was about eight, but even then I could see the horror and injustice of executing a teenager for a murder he didn't commit.' Bentley is buried in Croydon's Mitcham Road Cemetery alongside the Elephant Man's manager, Tom Norman, Frederick George Creed, inventor of the teleprinter, and Ronnie Corbett. Quite the gathering of Croydon alumni.

no one could have predicted, and it couldn't have been further from punk if it tried.

The concept for the Brit School came about after its founder Mark Featherstone-Witty watched the schmaltzy musical *Fame* and decided to create the real-life version in Croydon – a school that swapped out arithmetic for a capella, and chemistry for choreography. The entrepreneur Richard Branson got involved, and there was a glitzy roster of patrons, from Lenny Henry to Wayne Sleep. But the idea was not everyone's cup of tea, and some Croydonians were so incensed that the creation of the Brit School came at the expense of the east wing of the Selhurst Tertiary Centre on Selhurst Road that they took to the streets, chanting 'Shame on Fame!'

The 1,200 wide-eyed hopefuls who applied for one of the 300 places available when the Brit School opened in 1991 couldn't have cared less. They wanted in on this great dream factory experiment – an education that might rocket them far, far away from Croydon. 'There are more growing stars here than on the ceiling of a planetarium,' wrote the *Independent* journalist Kevin Braddock when he called in. 'The sight of a teenage boy in a leotard flexing hamstrings against a coffee machine is not uncommon, nor is there anything exceptional about a trio of theatre students improvising a Shakespearean tableau adjacent to the sandwich bar.'

This was a velvet revolution, or at least a leather-and-Lycra one; a different approach to doing things in the entertainment industry. At one point the prospectuses were shoutlined 'The Importance of Dreaming'. The Brit School's rise coincided with shows like *Pop Idol*, *Popstars: The Rivals* and *Fame Academy*, shows which fed into *The X Factor*, *Britain's Got Talent* and *The Voice*, although Stuart Worden, Principal of the Brit School and a man who has worked there for over three decades, is adamant that the Brit School did nothing to inspire such shows. Still, in some cases the crossover was plain for all to see: Leona Lewis, who won the third series of *The X Factor* in 2006, had actually learned her trade at the Brit School.

For this reason, it's easy to cast judgement on the Brit School in the same ruthless way reality TV has taught us to judge everything and everyone. But you can't escape the colossal impact it had – particularly on the landscape of twenty-first-century music. Stuart Worden's recollections of watching various stars of the future on their way up the travelator of fame and fortune are quite something:

> I remember Jessie J being amazing in *Sweet Charity* ... Katie Melua performing 'Spider's Web' at Fairfield Halls when she was seventeen ... Adele doing an early version of 'Daydreamer' in our theatre ... RAYE singing in a year 11 show at lunchtime ... Kate Nash performing as Hamlet ... Olivia Dean being in an ensemble show about changing the world ...'

The suburban Croydon campus also nurtured the talents of Imogen Heap, FKA Twigs, Kae Tempest, The Kooks, King Krule, Katy B ... Croydon may no longer be manufacturing its own cars, but the former motor town does have a whirring assembly line of pop stars.

Most luminous of them all was Amy Winehouse, a recalcitrant admirer of soulful music who'd wrestled with poor behaviour and likewise results at previous schools, but at the Brit School found popularity and purpose. While Camden is usually cited as the heartland of Winehouse's legacy, it was in Croydon that she sowed the seeds for her debut album, *Frank*. Vitally, Winehouse's Brit School tuition allowed her to keep her edge – even if that meant taking pot-shots at fellow Brit-schoolers: 'Hold the front page!' wrote the *Independent* in 2004: 'Katie Melua is "shit", according to stroppy north London anti-crooner, Amy Winehouse.'

In 2024 Croydon's roads teemed with double-deckers bearing advertisements for *Back to Black*, a Hollywood biopic charting Winehouse's rise and

CROYDONOPOLIS

fall. The film's director is Sam Taylor-Johnson, formerly also an artist of repute (she made the arty 107-minute film of David Beckham sleeping), and another Croydonian made good.

While the Brit School was still nascent, another avenue of music-making emerged on the pirate radio airwaves and in the clubs of Croydon – and it too had global designs.

Dubstep – a teeth-rattlingly bassy brand of syncopated beats – was born in a setting even more unlikely than the Brit School. In 1992 Big Apple Records opened as a small record shop next to the fruit and veg stalls of Surrey Street; in its first decade it became a salon for acolytes of progressive house, techno and drum & bass. It was somewhere for people to spin new discoveries to one another (Terry Leonard aka DJ Hatcha, who later went on to great acclaim, was brought in as the buyer), and make music together, often using PlayStations. A Big Apple Records label was launched – its logo, ingeniously, a banana peel – releasing tracks from a new guard of Croydon talent like Oliver Dene Jones (Skream) and Adegbenga Adejumo (Benga) and Digital Mystikz duo Mark Lawrence (Mala) and Dean Harris (Coki). Meanwhile, pirate radio station Rinse FM capitalised on the nearby Crystal Palace transmitter to bestrew London's airwaves with the sounds of dubstep, risking raids from the Department of Trade and Industry and ASBOs as it did.

It paid off. One Rinse listener was John Peel, who became particularly enamoured of Digital Mystikz, bigging them up as one of his favourite acts of 2004. Soon after this, Skream's 'Midnight Request Line' burst onto the scene, filling dance floors and becoming one of the all-time anthems of the genre. Meanwhile, across the road from Big Apple, in the Black Sheep bar, clubbers' pints of snakebite and black rippled to the phat bass lines, à la *Jurassic Park*, spun by resident DJ SGT Pokes, a barman-turned-dubstep legend. 'Black Sheep was the only place that was willing to get down and dirty,' Pokes later told *Time Out*. 'I remember Skream and Benga stage diving out of an eight-foot-high DJ box.' The Black Sheep was the Greyhound reborn to a different beat: Croydon had become Croydub, and the sound that had been simmering in Surrey Street for over a decade now really started to spread. Skream and Benga landed their own BBC Radio 1 show, and toured the globe with the intoxicating genre they'd helped to conceive. 'A new sound is infiltrating nightclubs across the world,' wrote the *Independent* in 2006. 'From humble beginnings in Croydon, its combination of heavy bass, minimal Detroit pulses and sombre rhythms has struck a chord with electronica fans.'

Something else was percolating too. In the early 2000s, grime, a garage-inspired strain of electronica which raced along at 140 bpm with frenetic emceeing over the top, surfaced in East London with the entrance of artists like Dizzee Rascal and Wiley. While punks had cobbled together a DIY movement out of shonky guitars and safety-pinned T-shirts, grime artists got their voices heard on pirate radio, their faces seen with freestyle videos shot on smartphones. Again Rinse FM was instrumental, socking it to the powers that be by pumping out tracks like 'Bonkers' that spoke for a disenfranchised generation. But just as the genre seemed to have reached its zenith and started petering out, it got a second wind. Except this was more of a tornado, and its name was Stormzy.

Born as Michael Omari in 1993, Stormzy was the man who picked up the grime baton, ran with it and didn't stop until he reached Glastonbury. But as one of the rizziest acts on the planet, he had surprisingly bookish beginnings. In a 2016 interview with the *Guardian*, Stormzy revealed that both his competitive nature and his love affair with words began at his local library in Thornton Heath. During the school holidays the young Michael would feverishly partake of the Book Trail reading scheme, awarded a badge for every book he read and reviewed. 'I would sit in the library all day, not 'cos I loved reading, just because I needed those badges,' he said. All the same, words clearly had a profound effect, because soon he was stirring his stories of an edgy kidulthood into his own cerebral brand of street lit.

Stormzy's rise was nothing short of cosmic. In 2014 his track 'Not That Deep' – the video filmed in IKEA Croydon's car park – whipped up waves, before the following year brought the diss track retort 'Shut Up', used to bring out boxer Anthony Joshua at the O2, and going on to shift over a million copies. Stormzy has scored three Brit and seven MOBO Awards, his own book imprint with Penguin Random House and – the ultimate accolade – batted off a Cliff Richard Christmas album to scoop the number-one spot ('Mistletoe and grime', ran one headline.)

More than with any Croydon musician before him, Croydon and its environs bleed into Stormzy's tunes and videos. 'WickedSkengman 4' raps of his younger days wandering through the Croydon 'ends' wearing a rucksack and sparking up joints. (The track also namechecks South Norwood's Caribbean eatery Bluejay Cafe.) Guesting on RAY BLK's 'My Hood', Stormzy conjures up a south London Gotham where you'd buy guns before you'd buy books. (Croydon, as it happened, had played the actual Gotham City a few years previous, in *The Dark Knight Rises*, in which Batman leaped out of a window on Wellesley Road.) You've also got to wonder if 'Not That Deep''s line about how it's not you who's bad but your area is a reference to CR0. Here was someone giving voice to a tranche of youth who were all too often underestimated and overlooked[*] – much like Croydon itself.

In 2019 Stormzy galvanised his superstar status when he became the first British rapper to headline Glastonbury's Pyramid Stage: a stupendous achievement to be joining a roll of honour that included acts like Springsteen, U2 and the Stones. Amid a rush of riot-esque fireballs and blue flashing

---

[*] Claims in 2016 that local police had been racially profiling Croydon's nightclubs by stopping the playing of bashment, grime and other similar music genres popular with BAME people were later upheld in an official report.

Grime artist Stormzy cut his teeth in Croydon, and by 2019 was headlining Glastonbury's Pyramid Stage in a Union Flag stab-proof vest designed by Banksy. *Alamy*

lights, he appeared in an instantly iconic Union Flag stab-proof vest[*] and, launching into 'Know Me From', made glorious history there and then. When he asked the crowd where they knew him from, w-where did they know him from, most of the 100,000 people in that Somerset field knew *exactly* where Stormzy was from.

---

[*] It was designed by Banksy, as a riff on the John Bull waistcoat. Later on in 2019 the vest appeared in a shop window on Croydon's Church Street, part of a fleeting Banksy pop-up displaying a 'range of impractical, bizarre and offensive merchandise'.

19

COSTA DEL CROYDON

No longer conjuring up office blocks like Martini bottles in a Tommy Cooper sketch, the Croydon of the 1980s and 1990s continued to prosper nonetheless. Between 1985 and 1993 over a thousand new businesses moved in. Manufacturing on Purley Way had eased up, but instead, warehouse-style outlets like MFI and Toys 'R' Us sprang up along it.

IKEA arrived in 1992. On opening day the Purley Way all but ground to a halt, and the queue of people waiting to get in were fed free Daims to keep them in good spirits. 'In hindsight it was a huge mistake,' Cathy tells me. 'Once you were inside it was impossible to find a way out ... I got stuck in there for about four hours!'.

Over 30 years on, IKEA's success is still evident from the two 300-foot brick chimneys banded at the top with the brand's trademark blue and yellow. 'I think the original idea was to put a restaurant up the top,' says IKEA Market Manager Gary Pearce, pointing at the two totems that mark the spot of the Swedish furniture store in Croydon's Valley Retail Park. Sipping a coffee in the Croydon IKEA canteen (it's the UK's busiest restaurant, I'm told, and sells three million meatballs a year), Gary explains to me that 'There was a discussion, I believe, just before we opened the store, to knock them down. The right choice was made to keep them because they are quite iconic.'

They certainly are. All that's left of the huge Croydon B Power Station which occupied this spot until the late 1980s, they still make quite the statement. 'What are those over there?' people will ask, as they sip their spendy cocktails in the Shard, and within seconds they're googling it and watching videos of 'Not That Deep', or the one of a base jumper parachuting from the top of one of Croydon's twin chimneys. It's a pity they were never converted into a revolving eatery to wheedle in out-of-towners, yet they continue to earn their keep (it costs £160,000 just to have them *surveyed*), acting as a beacon not just for IKEA but the whole of Croydon.

Cars still pour off the A23 into IKEA's car park, but now there are other ways of getting here. These days, IKEA Croydon brings in an average of 50,000 visitors a week, some to buy bedroom furniture, others just to eat cheaply in the restaurant, and around a third arrive by public transport.

In hindsight, of course, the entire power station should've been kept, then Croydon might have its own version of Battersea Power Station's über-posh shopping mall. True to form, Croydon was rash when it came to demolition. The same can be said for the nearby Purley Way Lido, whose closure in 1979 had signalled a last gasp of Croydon demolitionism. Towns all over Britain are now rediscovering, restoring and reopening their magnificent lidos. But once again, Croydon was so far ahead of the game it had long converted its own lido into a garden centre, another result of Croydon's retail park boom. The garden centre's gone too now, and the lone reminder of the lido now is the Grade II-listed, multi-levelled diving board, a graceful piece of Art Deco engineering.

At least to some extent, the lido was replaced by what was heralded as another beacon of Croydon's continued affluence: Croydon Water Palace. The largest indoor water park in Europe, it was opened (by the boxer Frank

The twin chimneys of IKEA can be seen from miles around, although sadly plans to convert them into a revolving restaurant didn't get off the ground. © *Will Noble*

Bruno, no less) opposite the old airport in 1990, bragging that 'Florida has come to Britain, palm trees and all!' You could almost believe it, the aqueous fun house bidding you to hurl yourself down six spiralling flumes, swerve down the 'Lazy River' in an inflatable ring and, as one promotional photo suggested, sit poolside in your budgie smugglers – hairy chest, gold chains and all – while slurping champagne with your better half. Live in Croydon, and you were living the dream.

This picture of rude health continued in the town centre. Croydon's Allders had grown into the third largest department store in the UK (though 1990's *The Return of Mr Bean* was filmed in the Allders in

CROYDONOPOLIS

All that's left of the Purley Way Lido is its Grade II-listed diving board. Often Croydon can move on too hastily for its own good. © *Will Noble*

nearby Sutton), and Croydon's Marks and Spencer was the second largest – only its flagship on London's Oxford Street near Selfridges was bigger. Throughout the 1980s and 1990s, the town's reputation as a shopping mecca continued; another shopping mall, the Drummond Centre, opened in 1986. Jugglers, musicians and clowns performed in the streets, bringing a touch of Covent Garden to Croydon, and fresh-faced businessman Tim Martin was vowing to rid the town of its 'shabby and loud' pubs with his new brand of hostelry, named after his old teacher, Mr Wetherspoon.

LGBTQ+ clubs and pubs thrived in the Croydon of the 1980s and 1990s. Queers partied in venues like the Bird-in Hand on Sydenham Road, PJs in Thornton Heath and Reflex in South End. 'Only twenty, thirty years ago, Divine was performing in Croydon!' says drag queen Shepherd's Bush, who co-runs the LGBTQ+ night Their Majesties in Croydon now. In 1993, Croydon hosted its own Pride festival, trumpeted by figures like Ray Harvey-Amer, a queer rights campaigner famous for his handbag which, says the Museum of Croydon, contained 'a dab of lippy for "fortitude against the oppressors", a whistle and bell to let the enemy know you were coming, gloves for glamour and a feather boa for joy.' It was also Harvey-Amer who inspired people to show solidarity with those dying from AIDS at Croydon's Mayday Hospital (then cruelly nicknamed 'May-die', and since re-christened Croydon University Hospital), which was being used as a terminus for young men at the time. 'He brought people to come and spend time with these men who were dying – their last hours,' says Shepherd's Bush.

The theatrical behemoth of the Fairfield Halls was counterpointed by the Warehouse Theatre, a small independent that'd started wooing Croydonians with lunchtime performances in 1977, and within ten years was helping lay the groundwork for Britain's alternative comedy explosion, hosting French & Saunders, Lenny Henry, Rik Mayall, Julian Clary and Ben Elton. Nearby, Croydon's Victorian town hall underwent a £30 million refresh; the new Clocktower complex featured a modern library, gallery and museum (Captain Sensible lent his voice to a cartoon crow that guided people through the exhibits), and the David Lean Cinema.

In an echo of the Marshall days, the Conservative council defied government-imposed limits on spending. The difference this time was that Croydonians and culture were at the forefront, not corporations and commercialism. Could it be that Croydon was *learning from its past mistakes?* Even Picasso got involved. In 1995, the Clocktower landed 80 of the Spanish artist's paintings, drawings, sculpture and ceramics for an exhibition – the sort of cultural coup that'd normally be left to the Tate or Royal Academy. Adverts for *A Picasso Bestiary* were pasted up on the Tube, suggesting with a self-detrimental, Skoda-esque wink that *Croydon itself* knew how far-fetched this all sounded: 'Picasso seen in Croydon – Yes, Croydon!'

It was the first years of the new century that saw perhaps the most high-profile articulation of ambivalence about Croydon, as well as the town's greatest on-screen moment.

Peep Show, the Channel 4 comedy wince-a-thon, first aired in 2002, and starred David Mitchell and Robert Webb as a brace of hapless housemates holed up in West Croydon. It started out as a modest, low-budget project, eyed with suspicion by its broadcaster – indeed, it's been said that Channel 4 didn't really 'get' the show and was keen to bin it. But thanks to a mushrooming cult following on DVD, along with the hearty support of Ricky Gervais, whose sitcom *The Office* had just redefined comedy, *Peep Show* became one of the best-loved Britcoms of all, running for nine series up until 2015.

Unusually for a sitcom it was shot on location, in and around Croydon. There are regular establishing shots of West Croydon's Zodiac House,[*] where Mark and Jez live (it also features in an early scene of series one, episode one, in which Mark is bullied into offering a group of kids a Coke, then accused of being 'a paedo'). For the first two series the interior scenes were even filmed in a flat in Zodiac House; watch the early shows and you'll

---

[*] Renamed in the show as Apollo House, which is confusing because Croydon has a real Apollo House too.

see real-life Croydon photobombing the scenes through the apartment windows.

Alas, Croydon didn't get quite the star turn it might have done – something that can be blamed squarely on . . . Ricky Gervais. Jeremy Wooding, who directed the first series, explains to me that he wanted Croydon to be an intrinsic part of *Peep Show*: 'For Mark to get on a tram every morning is just part of his odd life.' The execs weren't so sure. 'They were concerned that *The Office* had really pinned down Slough so much as being the place where it's set,' says Jeremy, 'and so they were very wary about saying "It's in Croydon" because it may have looked like we were copying *The Office*.' Croydon was forced to take a back seat, and this time round it had more reason to feel aggrieved at being elbowed out by Slough. 'The overall idea was that where they lived was Nowheresville and Anywheresville,' says Jeremy. 'It could be Basingstoke, it could be anywhere in that sort of southern commuter belty kind of way. I think at one point I said, "It's pointless not showing Croydon, because when this is a cult hit series, people are gonna want to know where it's set!"'

In the late 1970s there had been an outbreak of Croydonians wandering about in T-shirts that read 'Costa del Croydon'. These were sold by the *Croydon Advertiser* after it printed a story about the Croydon Chamber of Commerce angling to promote Croydon as a tourist resort with its own tourist information centre. Believe it or not, Croydon actually *got* a tourist information centre later on.[*] The whole Costa del Croydon schtick was only ever a bit of fun, but in due course *Peep Show* fans *did* find out where the show was set and, as the cult grew, a mini tourism boom took off, viewers trickling (if not flowing) into Croydon to gawp at Zodiac House, the outside

---

[*] There's a band from Devon called Croydon Tourist Office, following an episode in which a band member went to Croydon and asked the Visitor Centre how to get to the Museum of Croydon. They didn't realise there was a Museum of Croydon. When he finally found the museum, the people there apparently didn't know there was a Visitor Centre.

of JLB Credit where Mark Corrigan's grinding job plays out, and the Mexican restaurant where he pisses into a vat of jalapeño sauce. The 'Costa del Croydon' prophecy had come to pass – in a roundabout way.

And as *Peep Show* went from strength to strength, the writers beefed up Croydon's role, or at least made it less secretive. 'It kind of grew into its own Croydon-ness', admits Jeremy Wooding. In the ninth and final series there's a hilarious scene in which an unwanted housemate played by Tim Key is bundled up in a sleeping bag, waterboarded with lager and dragged into a lift. 'It's high jinks,' Mark assures a disturbed resident. 'We're the Croydon Bullingdon!'

But if you think Croydon got the *Peep Show* gig because of its 'punchline' status, you're mistaken. 'It was never chosen because it was a funny location,' Jeremy Wooding explains. 'It was chosen for practical reasons.' The prosaic truth is that the show's line producer Barry Read had solid connections in the area, plus lots of the cast and crew lived in either South London or Brighton. Again, Croydon's uniquely handy location came up trumps.

For all its Croydon links, *Peep Show* has a surprising connection to one Hollywood film. 'I suppose one of the big influences for me was the Jack Lemmon film *The Apartment*,' Jeremy tells me. In this skew-whiff take on Billy Wilder's classic 1961 romcom, Mark Corrigan is a spin on Lemmon's C.C. 'Bud' Baxter. 'The way he's a wage slave drone in a huge open-plan office, and how he's in hock to all his managers to try and get promotion, and therefore allows them to use his apartment for dates'. Meanwhile, the character of Sophie is the Croydon version of Baxter's infatuation, Fran Kubelik. Sophie was played by Olivia Colman, an actor who went on to Oscar-winning success with *The Favourite*. One half of *Peep Show*'s writing team, Sam Bain, then created one of the most acclaimed pieces of American TV in recent times, *Succession*. Croydon is used to acting as a stepping stone, but in this case a couple of people used it as a springboard to Hollywood.

*

Though the prophecy of the Tube reaching Croydon was never fulfilled, as the twentieth century neared its close, the town was bringing its transport A game. In 1992 the Victorian station at East Croydon was replaced with Alan Brookes Associates' ladder-masted glass box, which could cope with 10 million passengers a year. 'It's made of glass and soars into the sky,' gushed the *Surrey Mirror*, claiming the new station closely resembled 'some of the more spectacular European architecture in its styling', though failing to suggest *which*. (Less generous critics branded it the 'Oil Rig'.)

At the turn of the new millennium, something even more radical happened: Croydon got the transport infrastructure equivalent of a reverse vasectomy, by reinstating its trams. Stretching for 39 stations from Wimbledon in the west to Beckenham Junction, New Addington and Elmers End in the east, it was the only tram system in Britain south of the Midlands. In December 2000, a few months into the new service, a special royal tram, a royal pennant flag attached to its wing mirror, ran from East Croydon to New Addington, carrying on it the man who is now King Charles III. Once again, Croydon was luring in the royals.

At the heart of the Croydon Tramlink project was an idea to pump new life into the central retail district. 'There was a big boom in footfall in Croydon post-Tramlink, which was what it was intended to do,' says David Wickens, who worked as Head of Engineering and Project Management at Croydon Council during the time the trams were being reinstalled. 'It took cars off the road, it could get people in quickly, efficiently and effectively, whereas buses ... they're a bit prone to breakdowns and traffic congestion.' *Architects' Journal* was impressed: 'Croydon has quickly acquired the atmosphere of the tram cities of Holland and Germany,' it rhapsodised. Safe to say, Croydon believed its new trams buffed up its prospects of becoming a city in the future too.

At sun-up on a bright new millennium, Croydon, now with a burgeoning borough population of over 330,000, had regained its libido for assertive

Croydon's second generation of trams lend the town a touch of European glamour lacking elsewhere in London. © *Will Noble*

rebuilding, had a strong outlook on commercialism and culture, and boasted a new transport system that was the envy of the south of England. It was back at its go-getting finest.

What could possibly go wrong?

20

THE TOWN CENTRE CANNOT HOLD

The tram tracks at Reeves Corner were aflame. It was reminiscent of that scene in *Back to the Future* where the time-travelling DeLorean hits 88 miles per hour, leaving two fiery tyre tracks in its wake. Except this wasn't some zany Hollywood movie, nor was Croydon in film location mode. This was real life, and it was scary as hell. Croydon was burning to the ground, and the world watched on in horror.

Downfalls often aren't triggered by one major mishap, but a flurry of little ones, an insidious unpicking at threads that's almost intangible to begin with. Croydon's millennial optimism and prosperity didn't so much come crashing down as be very gradually plucked apart. Margaret Thatcher's 'Big Bang', deregulating the country's financial industry, twinned with the development of Canary Wharf in the mid-1980s, coaxed corporates away from the Croydon metropoffice. Companies now demanded shiny office blocks with cavernous, open-plan trading floors, air conditioning and underfloor ducting for the intranets and ultra-high-speed cabling necessary for digital money trades – and Croydon's Brutalist blocks struggled to offer all this.

The completion of the M25 in 1986 meant easy access to newer, bigger shopping experiences like the vast malls at Lakeside, which opened in Thurrock in 1990, then Bluewater in Kent in 1999. Closer to home, the Glades opened in Bromley in 1991, a posturing new contender to the Whitgift Centre's crown.

Meanwhile, the illustrious Grants department store had closed its doors for good. In due course the global phenomenon of Internet shopping (the unhinged offspring of the 'serve yourself' model pioneered in 1950s Croydon) threatened to siphon shoppers off the high street altogether. In 2000 and 2002 Croydon suffered the double humiliation of denial of city status. The second time, the ruling body had the brass neck to claim that Croydon had 'no particular identity of its own'.

The 2008 recession led to further fraying of the fabric. Plans for a huge new shopping mall, Park Place, were shelved in 2009. By April 2011 the council had shut down the David Lean Cinema. By now the town's music venues were petering out, too. The Fairfield Halls was muddling through with the tribute acts of the *actual* bands who'd played there in times gone by. The Beatles were replaced with the Mersey Beatles; Pink Floyd with the UK Pink Floyd Experience; Fleetwood Mac with Rumours of Fleetwood Mac; Bowie by the Bowie Experience.

Paul Talling is well-known for his *Derelict London* walks around the lesser-tramped parts of London, and on his Lost Music Venues tour of Croydon he strides through the streets and begrimed subways, a mini speaker dangling at his side blasting out Johnny Moped and the Damned. Every now and then he'll pause the group in its tracks at a venue that has either had its nightlife scooped clean out of it or been demolished altogether. After the Greyhound became the Blue Orchid in 1986 it was nicknamed the 'Blue Schoolkid', says Paul, because it was notorious for turning a blind eye to underage punters.

Recession, riots, austerity, mini-budgets, a Chinese property crash and various other problems have recently led to a slump in Croydon's fortunes. © *Will Noble*

That venue is long extinct too, although, in a sign of the times, you can now buy an *Orchidée Bleue Eau de Croydon* candle online, with a scent designed to 'take you back to the retro carpets and podiums which provided the backdrop to wild nights out at Blue Orchid.' Says Paul: 'I am not sure if I want that smell in my living room ...'

Croydon was beginning to unravel, but no one could have predicted the almighty thread-yank that came on 6 August 2011 – a moment when Croydon seemed to teeter on the verge of extinction.

'I can remember feeling the heat and seeing the flames from my bedroom window as a teenager and being absolutely petrified,' Hannah tells me. 'I then turned on the TV and saw on BBC News that a news helicopter was hovering over a house and said, "I recognise that house . . ." – then realised it was *my* house. Then someone jumped into our garden asking if we wanted a flat-screen TV.'

The House of Reeves furniture store, a business which had continuously traded for 144 years in the heart of Croydon and seen the road junction on which it stood come to be named Reeves Corner, found itself in the middle of this oafish lunacy. Protests over the Met Police's killing of a young Black British man called Mark Duggan up in Tottenham had quickly descended into something darker and destructive. A group of rioters set the shop alight and, as news cameras rolled, the place went up in flames. The conflagration was so violent that the nearby tram tracks caught alight, and a woman living across from Reeves had to hurl herself from an upper-storey window to avoid being burned alive. Reeves' Arts and Crafts showroom, with its glorious pitched roof, was so badly damaged bulldozers were called in to put it out of its misery. Eighty-year-old Maurice Reeves, who'd retired from the business, surveyed the scene of destruction and told the press that Shakespeare couldn't put into words how awful he felt.

Over in West Croydon, the Rockbottom music store, where so many musicians had come to buy guitars or simply chew the fat with owner Carl Neilson, was looted and gutted by fire, while petrol bombs were brainlessly hurled at flats along London Road, rendering people homeless within seconds. They simply sat out on the pavement wondering what to do next. Paul Talling went out with his camera the next day; a whole new Derelict London landscape had appeared overnight. What made this all the more jarring was that Croydon wasn't some inner-city project – a lot of people still thought of it as leafy Surrey. Though the rampaging was peppered throughout London and other cities in the UK, Croydon became an unsightly poster child.

Croydon has suffered its fair share of scandals, murders, accidents and tragedies. Forty-one locals died from typhoid in 1937 after drinking from a diseased well. Ten years later, one densely foggy morning, a young signalman called Horace Hillier mistakenly allowed a passenger train coming

The riots of 2011 saw the historic House of Reeves furniture store go up in flames, thrusting Croydon infamously into the national – and international – spotlight. *Alamy*

from Tattenham Corner onto the same tracks as a slower-moving train approaching Croydon from Haywards Heath. The first train careened into the back of the second at South Croydon Junction, killing 31 commuters and a driver. At the subsequent scene of tangled metal, 55 ambulances and 100 firefighters, Hillier was heard to mutter, 'It was all my fault.' In August 1961 the entire town was cloaked in grief after 34 pupils and two teachers from the Lanfranc School perished when their plane ploughed into a Norwegian mountainside. But the town always managed to keep its pecker up. As the journalist Tom Dyckhoff once put it, Croydon has risen from the dead more times than Nosferatu.

The 2011 riots were different. As smoke drifted balefully over the town, recalling the sooty gloom stoked by the colliers of centuries past, it signalled the culmination of a long, gradual period of decline, sometimes imperceptible, sometimes punctuated by big shocks, and the start of a period of stygian doldrums. What's more, everyone was watching. Among the newspapers that descended on Croydon as it burned, *Inside Croydon* noted, was the Italian tabloid *Corriere della Sera*. 'The riots exposed Croydon's shortcomings to the world,' a Croydon record shop owner told the BBC.

What the rioters had done was inexcusable, and yet it didn't come out of the blue. Over one in four children in the borough lived in poverty. As the local MP Andrew Pelling observed, the gap between the north and south of the borough had widened. Croydon's first food banks were just around the corner. As Krept and Konan, a rapper duo, half of which hails from Croydon, despaired in their song with George the Poet, 'Young Kingz Part 1': parents' debt could lead to a dead end wherever you tried to run …

Now things began falling to pieces. Withdrawal of funding saw the Warehouse Theatre closed in 2012. In a panic, the council then sold the family silver, or in this case £8 million-worth of the Chinese ceramics Raymond Riesco had gifted to the town, which prompted Croydon's museum to be stripped of its Arts Council Accreditation. Croydon, said the Director of the Museum Association, had been banished to 'the museum wilderness'.

The rest of the department stores went too. Allders was placed into administration in 2012, and closed for good shortly after. Kennards had become Debenhams in 1972, which closed in 2020 during Covid and never re-emerged.

Croydon's misery was compounded when Nestlé announced it was leaving, taking its 840 employees with it – headed for the bright lights of Crawley. It was like being unceremoniously dumped by a long-term partner. And if that didn't make Croydon want to sit in its pants and wallow in its own misery then 2012's third failed bid to become a city certainly did. ('Well, thank goodness that that piece of ill-advised, delusional nonsense is over,' sighed *Inside Croydon*, clearly tired of this civic windmill-tilting.)

What Croydon needed now was another inspired reinvention, a phoenix-like rise from the ashes. And as ever, it needed it fast.

If you were in Croydon town centre on 17 January 2013, you'd doubtlessly have spotted a gaggle of photographers surrounding a shambling figure crowned with a familiar shock of bleach-blond hair. It belonged to the then-Mayor of London Boris Johnson, and he was here to announce the relaunch that was going to hack through the town's Gordian knot. 'Croydon has huge potential to return to its former glory as one of London's most vibrant town centres,' boomed Johnson, with his usual over-abundance of confidence. He was referring to the announcement that Westfield, the shopping megamall chain that had already enjoyed huge success in White City and Stratford, was coming to Croydon next. The developer Hammerson, which already owned the Centrale Shopping Centre, would also take over the site of the Whitgift Centre, transforming the two into a £1 billion Westfield complex with 1.5 million square feet of retail space. The scheme promised 5,000 new jobs and 600 new homes, as well as putting Croydon back where it belonged – as one of the UK's top 10 retail and leisure destinations. It would be the envy of South London. The Whitgift Foundation duly started debouching traders from the Whitgift Centre.

There was more promising news to come. In 2015, in a bid to bring much-needed new housing projects to the borough in the face of a lack of central government funding, the then-Labour Croydon Council teamed up with a new developer called Brick by Brick. In return for a council loan of £200 million, Brick by Brick would deliver stylish, affordable homes conceived by top architects, including RIBA Stirling Prize winner Mikhail Riches. Many of these homes would be built on infill sites owned by the council across Croydon – a savvy use of land, with profits pumped back into the local economy. 'Built in Croydon, for Croydon' ran the Brick by Brick strapline. Croydon was leading the way in forward-thinking regeneration, taking the housing crisis into its own hands. The developer would also deliver the much-needed refurbishment of Fairfield Halls.

Though the Fairfield had hared off into the lead in the 1960s, it had gradually been overtaken by hungry rivals. The Churchill Theatre opened in Bromley in 1977, as a neighbouring competitor to the Ashcroft. The Barbican Centre showed up in 1982, lavishly subsidised by the City of London, its concert hall able to seat around 150 more people than the Fairfield's, and a lot more central. In 1996 the Royal Albert Hall's notorious swimming-pool acoustics were vastly improved, making it infinitely more attractive to orchestras and rock stars alike. In 2003, the Ambassador Theatre Group took over the reins of Wimbledon Theatre, which enabled it to begin putting on touring productions of major West End Shows. The Fairfield, meanwhile, had suffered from the departure of the London Symphony Orchestra, who'd found a permanent home at the Barbican, and even the decline of British wrestling thanks to the advent of the glitzy World Wrestling Federation, not to mention years of financial tension. But Brick by Brick was now promising to return Croydon's pride and joy to its 1960s/70s glory days.

THE TOWN CENTRE CANNOT HOLD

As for the question of what to do with St George's House now Nestlé had cut loose, that had been solved too. It was to be turned into luxury apartments for ritzy young professionals, and rebranded as Highgrove Tower, part of a sweeping regeneration project that would see the St George's Walk shopping centre converted into an array of apartments collectively known as Queen's Square. The next big Croydon facelift was on.

But while Brick by Brick and the council got stuck into the work at hand, Croydon was about to be bludgeoned with more grotty luck. In the winter of 2015, a handful of cats started going missing around the Croydon area. Soon it became clear they hadn't just got lost, they'd been killed – decapitated, in fact. Raw chicken was found in the stomachs of some; a maniac was creeping about in the dead of night, luring in people's pets, then ripping their heads off.

The national press fell into a frenzy, splashing headlines like 'Jack the Ripurr'. Panic set in: was the culprit just getting warmed up? Would they start attacking *people* next? Not only were Croydonians afraid to let their cats out, they were also anxious about leaving the house themselves. The actor Martin Clunes, an erstwhile Croydonian, labelled the attacks 'the stuff of nightmares', and the Met Police launched Operation Takahe, in a bid to winkle out the pussycat-killing psychopath. The killings continued – dozens, then hundreds. Murdered moggies started showing up in Surrey, Kent and various other parts of London. Either the Croydon Cat Killer had expanded their ring of terror, or copy Cat Killers were on the prowl. The murderer was rebranded as the 'M25 Cat Killer', yet would always be remembered for starting out their twisted games in Croydon.[*] And in the

---

[*] A £130,000 investigation by the Met Police concluded that the Croydon Cat Killer didn't exist. It had been foxes all along. Many Croydonians are not so sure. Boudicca Rising from the South London Animal Investigation Network (SLAIN) tells me that at least dozens of cats in Croydon were, without a doubt, killed at the hands of humans. Night patrols are still being carried out in the hope of catching the guilty parties.

midst of the investigation, Croydon was about to face up to a killer no-one had anticipated.

> Oh mate ... 30 of us on the tram this morning and we all thought our time was up ... tram driver took the hard corner to Sandilands at 40mph!! I swear the tram lifted onto one side. Everyone still shaking ...it's mad.

This is what a passenger on a Croydon tram posted on Facebook on 31 October 2016. It is 15 times safer to travel by tram than by bus, which itself is 24 times safer than travelling by car. But that Facebook post was a warning shot that suggested London's trams might not be quite as safe as most people thought. It didn't take long for these concerns to be confirmed. Just nine days later, on the same ashen November morning that newly-elected Donald Trump vowed to be a 'president for all Americans', reports started coming in of a tram accident near the Sandilands tunnel just outside central Croydon.

On Easter Monday 1907, a tram chock-full with passengers running between Croydon and Sutton had overturned on a bend at Park Lane in Carshalton, killing two and injuring dozens more. This time the consequences were worse: seven dead and 62 injured. Bleary-eyed Brits who'd stayed up all night to watch the election results trickle in were suddenly staring at live footage of a tram flipped on its side, surrounded by stunned emergency teams. The tram had been travelling at up to three times its 20 kph speed limit when it derailed, and survivors of the crash later described it as 'like being flung about inside a washing machine'.

The driver was ultimately found not guilty of failing to take reasonable care of his passengers, but Croydon was shaken to the core. The following year, its trams were fitted with an infrared safety system: if a driver closes their eyes, or takes their focus off the tracks, for more than a couple of seconds, an alarm starts beeping and the driver's seat rumbles like a gaming chair. But the damage was done. The ninth of November 2016 marked Britain's first fatal tram crash in decades, and again Croydon was making unappealing headlines.

While the United States had been blindsided by Trump's victory in 2016, the UK had also given itself a shock by voting for Brexit. When mixed with the 2008 crash, government austerity, Covid, a Chinese property crash, Russia's war on Ukraine and the ensuing energy crisis, and Liz Truss's 'mini budget' – not to mention misjudgements from various Croydon bigwigs – the result was pure poison. 'The country has suffered, London has suffered. But Croydon is a complex case,' Steven Downes from *Inside Croydon*, a website that tirelessly covers Croydon and its politics, tells me.

> It's part of London, of course, and has many of the problems of an inner London borough, but only gets funding for an outer London borough. That compounds the issues. No other outer London borough has a Home Office immigration office in the town centre, though, and Croydon's council has never been provided with the resources to cope properly with this national role that it has.

Hammerson had been showing signs of cold feet about Croydon's Westfield development for a while, and in the hardships of the pandemic in 2021 brought the project to a stuttering halt. Meanwhile, workers downed tools on St George's House after the development was hit by debts resulting from the Chinese property crash.

The tower was left shrouded in a skeleton of scaffolding, while the flattened St George's Walk shopping centre remained reminiscent of a Second World War bombsite. At the time of writing the online brochure for the binned-off Queen's Square was still reading (and this is genuine):

Yet more text. And more text. And more text.

And more text. And more text. And more text. And more text. And more

text. Oh, how boring typing this stuff. But not as boring as watching paint dry. And more text. And more text. And more text. And more text.

Boring. More, a little more text. The end, and just as well.

And then the sucker punch. Brick by Brick imploded. The plan, it transpired, had been full of holes. The complexity of many of the chosen sites had been underestimated; progress was slow, and in many cases went excruciatingly over budget. Some of the mistakes were laughable. With homes that *were* built, it emerged that Brick by Brick hadn't registered itself as an approved shared-ownership supplier, so couldn't legally *sell* them. After five years in operation, Brick by Brick hadn't paid a penny back to Croydon Council. A report later showed the interaction between Brick by Brick and the council had been 'over-reliant on relationships rather than formal structure'. In July 2021, the council, now with a gaping wound in its finances, decided to wind Brick by Brick down. But Croydonians had seen the last of that £200 million loan. What made this all the more painful was that the wait time for social housing in Croydon was around *10 years*. Around the corner from Brick by Brick's George Street's headquarters homeless people were sleeping in tents or on mattresses.

As for Brick by Brick's Fairfield Halls renovation, that had blown up into another scandal. Delivered a year late, it was also almost £40 million over the original £30 million budget, prompting a probe into whether fraud had been at play. 'The post-renovation has been a disaster,' classical music critic Norman Lebrecht tells me: 'fifth-rate visiting orchestras, poor programming, lack of self-integration in the London music scene; complete lack of strategy'.

The rotten cherry on this unsavoury cake: in November 2020, Croydon

St George's House was due to be turned into luxury apartments, but the project was put on ice during some of the toughest years Croydon has endured. © Will Noble

Council issued a Section 114 notice, effectively declaring itself bankrupt. As per procedure, it would do the same again in 2021 and 2022. Croydon was sinking.

On a mizzly March morning in 2023, the borough was putting on a brave face. The setting was Boxpark, and the soundtrack was a specially commissioned piece called *Oratorio of Hope*, created by Croydon's Grammy-nominated Tarik O'Regan, alongside a slew of local musicians and arts groups. It heralded Croydon's coronation as London Borough of Culture, the beginning of 12 months of cultural celebrations. 'Could this be a new start for Croydon?' *The Times* asked, as sculptures of giraffes were scattered around town under the slogan 'Croydon Stands Tall'. The sense of optimism in Boxpark was so thick among the assembled locals and VIPs, they could've scooped it up and spread it on their complimentary croissants.

But something didn't sit right. Just five days earlier, Croydon's newish Mayor, Jason Perry, had signed off an unprecedented council tax hike of 15 per cent. Croydonians in the midst of a cost-of-living crisis were forced to stump up even more cash – despite fearing that most of it would be lobbed into a financial black hole. Posters slapped up on lamp posts and bins pictured Mayor Perry dripping in bling like some browbeating gangster. Two speech bubbles spouted demandingly from his mouth: '15%', 'NOW YEAH!!'

As 2023 ploughed on, it became an annus horribilis for Croydon, one catastrophe chasing another. Stabbings in the town centre; reports of 'secret Chinese police stations' operating in plain sight;[*] *Private Eye* accolades for 'the ultimate rotten borough'. It was also a year for goodbyes. The huge Sainsbury's that had been part of the Whitgift Centre since 1969 closed. Newly minted ventures barely had time to get going. The Selsdon Park Hotel[†] was revamped as the bougie Birch country hotel, but had only just flaunted its rewilded fairways, gorgeous lido and Jay Rayner-applauded menu when it was liquidated.

An even more impactful closure was announced. Old Palace School announced it would close down in 2025. The Whitgift Foundation's latest flirtation with property speculation, and the fact that most of the Whitgift Centre's retailers had been emptied out to make way for Westfield, were beginning to bite. *Inside Croydon* wrote of a drop in the Foundation's annual rental income from £5.6 million in 2021 to £1.2 million in 2022. Covid had played its part of course, but it was more than that alone. Ever since the Sisters of Mercy had shown up to bail

---

[*] Though charges for this were later dropped.
[†] The same rambling country pile in which Edward Heath had hatched his 'Selsdon Man' plan, and where football teams often stayed the night before playing in the FA Cup final.

out Croydon Palace in 1887, the Old Palace had been run as a school, but that was now coming to an end. Students stood outside the Whitgift Almshouses with placards: 'JOHN WHITGIFT FOUNDATION HAVE RUINED CROYDON.' That night, if you'd stood in the Minster next to John Whitgift's tomb, you could have heard the distinct sounds of turning.

This bolt from the blue was still being digested when something truly horrific happened. On 27 September 2023, 15-year-old Elianne Andam, a much-loved pupil of that same Old Palace School, was fatally stabbed outside the Whitgift Centre as she waited for a bus.

Stabbings were nothing new to Croydon: in 2021 the borough saw five teenagers knifed to death, earning it the unenviable title of London's 'knife crime capital'. But Elianne's death touched a nerve in each and every Croydonian. It was the sheer callousness of the attack; that Elianne was so young; the suggestion she'd simply intervened to protect her friend when an ex-boyfriend became confrontational. As the Wellesley Road closed to traffic and the pavement by the Whitgift spilled over with flowers and votive candles, it was as if the town itself had fallen into a silent stupor. Croydon wasn't 'standing tall' like one of those painted giraffes. It was skulking in the corner, head bowed.

With a collective sigh, Croydonians asked themselves: *where the hell do we go from here?*

21

CROYDONOPOLIS

In 1995, a full-page editorial was published in the *Croydon Advertiser* informing readers that the resplendent Victorian water tower in Park Hill was to be turned into a helter-skelter. 'It's a real cute little olde-world building,' purred the corporation's Vice President Hank Ramrod to the *Advertiser*. 'OK, we'll need to tear out most of the insides, but the exterior will stay exactly the same except for the slide, a few lights and a burger bar ... I really can't see what the fuss is about, it's over a hundred years old anyway.' The developer behind this? The 'Total Fantasy Corporation'. The piece was, naturally, published on 1 April.

Every Croydonian who read this would have either laughed at the deft lampooning of their restless, reckless town – or taken the bait and swallowed every word. Smashing it up and starting again is in Croydon's blood. It is incorrigibly fidgety – impatient to move on and find the next panacea. If Croydon were a person, you'd find them in one of those grotty casinos on the North End: one more roll of the dice, bet the house, go big or go home (if you've still got one). It's a mentality that often pays off, in the short term at least. Think of the Surrey Iron Railway, the Croydon Canal, the Atmospheric

CROYDONOPOLIS

Railway, the airport, Croydonisation, the Water Palace,[*] Brick by Brick – all grand designs that started out looking like the Best Idea in the World. But one after another they crashed and burned, some with alarming speed. The truth is, if you're willing to act so impetuously, you must take the rough with the smooth.

But isn't this scorched earth, castles-in-the-air approach somehow familiar? For the atmospheric railway, substitute the *Titanic* or *Concorde* or the highly influential yet ultimately doomed Advanced Passenger Train. For Croydonisation, switch in the post-war rebuilding of Glasgow or T. Dan Smith's stab at turning Newcastle into the 'Brasilia of the North'. For the ill-fated strivings of the cash-starved council and Brick by Brick, think Warrington Borough Council's abandoned solar farm, Thurrock Council's series of failed investments or Woking's spiralling debt after hurling itself into property development.

Isn't Britain like Croydon – endlessly hungry to produce idealistic things, only to have them go on the blink or conk out altogether, either because they weren't thought through properly, weren't maintained, or were simply far too quixotic in the first instance? Look at Brexit: the most over-hasty, baby-out-with-the-bathwater stunt attempted in the country's modern history.

Croydon, in its own screwball way, is a microcosm of Britain. Its diversity of denizens, its mixture of urban grittiness and rural beauty, its antipodal wealth and poverty, its amazing cultural fecundity. But most of all, its 'all-in' attitude. Indeed, Britain has often found itself looking to Croydon as a litmus paper, its airports, racing tracks, business districts, tram systems all informed by the Valley of the Crocuses' triumphs and failures. And look at that night in June 2019 when Stormzy spoke for an entire generation. Wherever else in the country you are, don't look down on Croydon; you're living in one.

*

---

[*] This closed just six years after opening, haemorrhaging cash and plagued with reports of health and safety concerns.

Green shoots have already started to poke through the smouldering ashes of the riots. 'It was the riots that triggered us,' Mark Russell, co-founder and head brewer at the Cronx Brewery, which pours pints of Cronx Lager at its trendy High Street taproom, tells me: 'to go, "Well, let's do it – let's do something good in the community. Let's bring something back to Croydon, which in our case was a brewing industry."' In Victorian times, Croydon had become a brewing powerhouse, thanks to major breweries like Page & Overton and Nalder and Collyer's. The latter was known for a porter called Entire, so Cronx now replicates it. 'We sold our first beer about a year after the riots and went from there,' says Mark. Croydonians have lived and grown stronger thanks to the cloud of underdog-ness beneath which they reside. Says Croydon's Bishop Rosemarie: 'For people who live in and around Croydon, there can be a sense of "us against the world", of, "People think our town is crap but, hey ho, we love it and we're proud of who we are."'

Borough of Culture was, in many respects, a success for Croydon, a cathartic and sorely needed yell of 'Look at what we've achieved!' The Twentieth-Century Society did tours of the most endearing of the 1950s–70s high-rises. The David Lean Cinema held a festival dedicated to Croydon's punk scene. Pollock's Toy Shop, jettisoned from its permanent Fitzrovia home, sought solace with a pop-up in the Whitgift Centre. Plaques were studded into the pavements for Kirsty MacColl, Desmond Dekker, Big Apple and many others. Young people engaged in the power of possibility with gaming workshops from Lives Not Knives, and a Young Croydon Composers showcase. Brit School students produced a musical on the theme of climate change. There were life-affirming performances from D/deaf and disabled artists during the Liberty Festival. Chineke! performers joined with young musicians from Croydon Youth Orchestra and Trinity Music Academy, to play Gershwin, Sibelius and, of course, Coleridge-Taylor.

Outside of Borough of Culture, things were happening too. In the run-up to the 2024 Oscars, Croydon was unexpectedly shifted into the spotlight, thanks to Andrew Haigh's nostalgic ghost story, *All of Us Strangers*. Many scenes were shot in Haigh's childhood home in Sanderstead, while the Whitgift Centre was briefly returned to past glories in a tear-jerking denouement. The *LA Times* named it the best movie of 2023, and the film's success was enough for the *Guardian to* go as far as heralding 'the unloved south London borough' Croydon as suddenly 'hot property'.

A few days before 2024's Oscars, the Brit School graduate and ardent Croydonian RAYE scooped a record-shattering *six* Brit Awards in one fell swoop, doubling Stormzy's all-time tally in a matter of hours. As she took to the stage for the umpteenth time of the night, she roared into the microphone so all the world could hear: 'SHOUT OUT TO CROYDON, SOUTH LONDON!'

All these things hinted at a promising hereafter for Croydon. Yet some remain pessimistic about any kind of future. 'Croydon is a sad town that seems to have been left to disintegrate,' says Peter Fox, the punk DJ. 'Croydon's gone into the doldrums now,' David Wickens, who worked developing the Croydon trams in the 1990s, tells me. 'I think Tramlink was the last throw of the dice to get something done, and they were just too late.' Once the envy of those Balhams, Brixtons, Peckhams, Streathams, the town now seems to be playing an uphill game of catch-up. On Facebook groups recalling the good old days, words like 'sad', 'unrecognisable', 'run-down', 'shame', 'breaks my heart' and 'ghost town' are spat into a cyber-spittoon. Everything, according to some former Croydonians, has gone to hell in a handcart.

You can see where they're coming from. After all, they got to bathe in the rays of one of Croydon's golden eras. Luxury lunches in the top-floor department-store restaurant, the rampant job opportunities, the ear-splitting rock bands, the top orchestras, the plenteous department stores, maybe even one of the post-war flights from the airport. How could they be expected to be invested in a place that no longer has any of these?

Perversely, in recent times, perhaps Croydon has suffered from *too much* optimism. In 1993 the BBC broadcast an edition of its cultural review programme *The Late Show* from Croydon. The programme started, a bit like this book, as a litany of assaults against the town, spliced with clips from the futuristic French film noir *Alphaville*, in a nod to its 'English Alphaville' moniker. The highfalutin' Pompidou Centre architect Richard Rogers purred that Croydon 'certainly isn't somewhere I'm going to spend my summer holiday in, let alone my weekends' (surely he meant that the other way round?). This wasn't just a celebrity roast: *The Late Show* was covering a new exhibition, *Croydon the Future*, hosted in a gazebo beside the Fairfield Halls and inviting Croydonians to pore over high-concept plans from 14 edgy architects for transforming the town.

In this Babylonian Croydon, pedestrians would float over dual carriageways on sculptural, boomerang-shaped bridges, and wander through a linear art gallery and sculpture courts running the length of Wellesley Road. The tops of car parks were piggy-backed by inflatable domes, giant plastic marshmallows hosting concerts, showjumping and basketball games – even ski runs with real snow. Rogers himself wanted to implant a 150-metre-tall spire, not dissimilar to the Skylon from 1951's Festival of Britain, or indeed the 'Tower of Light' proposed for Croydon a few years later, with electricity-generating turbines.

'Will this brave new Croydon ever be built, or is it just a PR showcase for architectural pipe dreams?' asked the *Late Show* presenter. The clue was in some of the on-screen presentations. 'People could arrive from East Croydon Station,' explained Ian Ritchie, wearing a salmon-pink sports jacket, who'd suggested burying Lunar House beneath a huge ornamental lake, 'take a boat, and actually have a most extraordinary paradise of a wait to get themselves registered or deregistered or whatever they do at Lunar House.' It wasn't a good omen that the exhibition display tent was torn to shreds in high winds that December, badly damaging the £40,000 futurama of a Croydon that was never meant to be.

It was just one of various Croydons that haven't seen the light of day. In 2007 the avant-garde architect Will Alsop presented a masterplan for a Croydon of tomorrow that would propel it into the limelight as London's 'Third City', alongside the City of London and the City of Westminster. 'Croydon needs to dare to dream – it should set its sights high,' Alsop proclaimed, as he revealed his design for a 'vertical Kew Gardens' in the centre of the town – a latter-day Beulah Spa. A *Sunday Times* journalist found Alsop chain-smoking with a scattering of plans of this Brave New Croydon in front of him, plus two bottles of wine. 'The developers are gathering like bees round a honeypot ...' Alsop told the journalist. 'Croydon is pregnant with opportunity.' Meanwhile, Ken Shuttleworth (who'd worked on London's Gherkin) planned to build four skyscrapers by East Croydon Station that looked like cut diamonds. It was a statement, certainly – but was it one that belonged in Croydon?

Since the mid-twentieth century there has been a disconnect between Croydon and the people physically reshaping it. How to harmonise a striking, paradigm-shifting reinvention with the selfless imperative to simply improve the daily lives of its citizens? In the Sixties and Seventies there was no cohesive plan. By the 1990s the town had become a tempting pinboard for starchitects' ego-driven schemes. Many developers still see Croydon as beneath them. In his book *Iconicon*, John Grindrod reveals that the attitude of post-millennium developers could often be: 'Look, it's only Croydon. We don't use our best teams for Croydon.' 'Croydon was a place that was very good at coming up with big, visionary plans and ideas,' adds Vincent Lacovara, 'but not so good at making them happen.' Will

Alsop's Third City never materialised either. He died in 2018, without a single foundation dug.

But Croydon is always going to build *something*. Despite the abject failures of the Brick by Brick scheme, Westfield and Queen's Square, not to mention several futuristic mirages courtesy of out-of-touch planners, the town continues to build heartily. Yes, it sputters and stalls – that's evident in the perma-scaffolding worn by some of the buildings. But recent years have given Croydon the 43-storey Saffron Tower, its purple-pink patchwork cladding a rather literal nod to Croydon's 'Valley of the Crocuses'; Ten Degrees, the 44-storey-tall block which, at the time of completion in 2020, claimed to be the world's tallest modular building; and its jagged-topped neighbour, Enclave, with its 'Cosy Studio' floorspace sizing up to just 218 square-feet. It's a very different style of living to High Towers, that rambling Park Hill villa advertised in 1903.

The burgeoning of Croydon's population may have abated since those times, but these towers are monuments to the fact that the town alone has some 200,000 residents, with almost double that across the entire borough. By 2031, the borough population is expected to be not far off half a million. Up is the only way for Croydon, and that's not going to change anytime soon. Indeed, in 2023 a blocky version of the Mayor of London Sadiq Khan appeared online, inviting kids to take part in a competition to design a 'Croydon of the future' using Minecraft.[*] The next generation of people who will reinvent Croydon yet again is already being sourced.

While unable to kick its habit of getting sky-high, Croydon has at least learned a lesson or two from its days of savage sacrifice-making. Though the Roundshaw housing estate and Wilson's School nowadays occupy much of its green space, the majority of Croydon's airport has mercifully avoided

---

[*] All the more fitting because some of Croydon's latest buildings look as though they were built on Minecraft.

demolition. Today Airport House, as it's now known, is a business centre, but much of that trailblazing airport building from 1928, including its glass-domed atrium, where the gentleman in spectacles, spats and an Ulster had once waited with his wife, is still accessible to the public. Granted, you can no longer catch an Imperial Airways flight to Karachi, but you can get a decent chicken karahi from the Imperial Lounge and Restaurant nestled in one corner. You can also still order a cocktail at the Aerodrome Hotel, though your chances of running into a pilot are slim.

Croydon Airport, fronted by a plinth-mounted de Havilland DH114 Heron, the same model of plane in which Geoffrey Last flew that valedictory flight back in 1959, opens to visitors once a month. A group of volunteers take you up into the control tower, now a reliquary stuffed with fascinating trinkets, from vintage booklets informing female passengers how to fly, to the wicker chairs once bolted down to early aircraft. In one glass case is a caramel-coloured flight bag with two leather stripes down the centre. This is what Amy Johnson had with her on her final, fateful flight. 'It's a bit like the Elgin Marbles,' you might hear a tour guide say.

While its airborne heritage is a swelling source of pride for Croydonians, so too are the trams of today. In the last decade there have been around 250 million tram journeys made on the South London network. The trams are also a rare instance of having something the rest of London – and most of Britain – still can't. There's a hunger in London to bring trams back on a wider scale, but since its original tram network was paved over, many things have been put beneath the streets: drains, electricity, telecoms. Digging them up would be an unmitigated headache, which is why so many mooted tram schemes, including one along Oxford Street, have never taken off. 'The big advantage of the Croydon tram system,' says David Wickens, 'is that a lot of it was built on old railway track, and virgin land.'

More than that, trams have become an icon of Croydon and its people. Find yourself at the Arena tram stop on a Saturday afternoon and you'll

Miraculously perhaps, much of Croydon Airport still survives, and is now home to a museum crammed with aeronautical memorabilia. © *Will Noble*

often hear the animated chant of 'Come on you Trams!' wafting across the platform from the diehard supporters of Croydon FC, the town's Southern Counties East League Division One football team.[*] As the final whistle blows at each match, Trams team members jog over to their fans, and shake the hand of each and every one.

---

[*] On 17 February 2024, The Trams' Ryan Hall struck the ball from the halfway line at kick-off against Cockfosters Reserves, and it hit the back of the net 2.31 seconds later, in what was (unofficially) the fastest goal ever scored in football.

CROYDONOPOLIS

Such spirited support of Croydon FC hints at something else that's underrated in Croydon – something found hiding between the gaps of the high-rises – a sense of pride and community. 'What I most love about Croydon is that it is an incredible melting pot of people with different cultures, backgrounds and experiences,' Shaniqua Benjamin, Croydon's Poet Laureate, tells me.

> You never know what accent or language you'll hear when stepping out the door. You smell aromas of food from across the world on one street. You've got the aunty talking to the mandem on the high street. I'm just in love with the mixed bag of nuts that Croydon is. For all those who see Croydon in a negative light, I'd say there is so much beauty grown from the struggle that has led to a real strong community spirit, which isn't always allowed to shine. The greatness is there if you look for it.

Community in Croydon flows as deep as the concreted-over River Wandle. You'll find it in the meet-ups of the Croydon Natural History and Scientific Society, where discussions about Zeppelin attacks and wildlife around the old aerodrome are held in a parochial hall sitting in the shadow of One Croydon. On Surrey Street Market, where the cheery vendor Johnny Herbert has been working since he was six, and continues to sell tomatoes and bananas with a grin. It's in the Queen's Gardens every night of the week, as Nightwatch volunteers hand out soup, sandwiches and sleeping bags to Croydonians unfortunate enough to be homeless. On Norwood JunkActions's weekly litter picks, where the detritus is recycled into art installations like beer can trees and replica clock towers. In fact, once you open your eyes to it, a torrent of pro-Croydon set-ups becomes evident: Croydonites, Croytopia, Restart Croydon, Reclaim Croydon, Check Out Croydon, Young Croydon, Our Space Croydon, East Croydon Cool, Croydonist . . .

The Oval Tavern hosts LGBTQ+ cabaret night Their Majesties, where hosts and co-founders Shepherd's Bush and Asifa Lahore (the UK's first out Muslim drag queen) have created a loving community, gleefully reigniting the one that existed decades before.

Every Christmas, Their Majesties put on a Croydon-themed nativity, in which queers get to re-live their school nativity and put a positive stamp on it. Asifa tells me: 'For a lot of people there is a sense of isolation in Croydon, because obviously in terms of land we are one of the largest boroughs ... We find people coming up to us and saying, "Oh, my God, this is amazing!"' Adds Shepherd's Bush: 'We've had people from Bristol, Manchester, Doncaster who've made the trip to come and see little old us ... We've had people from Miami!'

LGBTQ+ club night and safe space Their Majesties has made Croydon a hotspot for the queer community: 'We've had people from Bristol, Manchester, Doncaster ... Miami!' © *Will Noble*

CROYDONOPOLIS

Despite going stratospheric, Stormzy remains in tune with his hometown, echoing the way in which Samuel Coleridge-Taylor spoke up for his fellow man and woman. The rapper's #Merky Foundation charity throws free Christmas parties for local kids at the Fairfield Halls. In 2023, Stormzy announced that, along with the Crystal Palace striker Wilfried Zaha, he was buying non-league club AFC Croydon. Give it time and perhaps Croydon will have two Premier League football teams. In the words of *NME*: 'Stormzy doesn't forget Croydon, and Croydon doesn't forget Stormzy.' Indeed it doesn't: you'll find a plaque dedicated to him outside Thornton Heath Library. A statue is surely only a matter of time; maybe they'll put it next to Samuel Coleridge-Taylor's.

The Whitgift Centre – that much-maligned kernel of Croydon where this journey started out – is also home to pockets of love for the Cronx. Among the Magic Corn stalls and outlets selling diamante-encrusted bed frames, you will find the Windrush Generation Legacy Association, with its 'Gaan a Farin' replica sitting room, and regular community meet-ups over coffee and plates of saltfish and ackee.

Turf Projects and the Croydon Urban Room are two places where locals can congregate to sketch out ways in which they'd change Croydon for the better and pin up their thoughts and hopes for its future. Then there's Bishop's Wine Bar, a burgundy-walled time capsule stuck firmly in the late 1970s/early 1980s when it opened (no-one seems sure of the exact year), frequented by characters like Inky Dave and Sweary Brian, and run by Angela. 'If I had a pound for every time someone said, "I've lived in Croydon all my life and I didn't know you were here," I wouldn't be stood here,' grins Angela from behind the bar. 'I'd be in the Maldives.' Like the Minster, the palaces and Airport House, Bishop's is another time capsule of Croydon, proof that the town never did away with *everything* from its many golden eras.

Croydon will always be about going large: infrastructure firsts, boomtown buildings, city statuses, corporate giants, Hollywood big shots, world records, chart-smashing hit-makers. Yet so much of what makes it great derives from seemingly humble gestures, like Sislin Fay Allen applying for that police role or Jamie Reid whipping out his pen to sketch some behemoth bogeymen. For all that the church, council, corporates and cartographers have done to build and bully Croydon into the shapes

'Gaan a Farin', a replica Windrushers' sitting room at the Windrush Generation Legacy Association, one of a number of community gems stashed away in the Whitgift Centre. © *Will Noble*

and structures they desire, it's the low-lying grassroots movements from which the next RAYE or Roy Hodgson, Jane Drew or David Lean, will emerge. As Nadia Rose tells me, 'We'll *always* be the borough of culture.'

Dismiss Croydon as a punchline and the joke's on you. As the local historian John Gent once wrote, 'Croydon is often maligned, mainly by those who know nothing of it.' People will always have bad things to say about Croydon. As the Luftwaffe demonstrated, it's an easy target. But the truth is, there is a bit of Croydon in wherever we live, and a little of Croydon inside us all. Which is why everyone should come here at least once.

Let's reinstate that visitor centre. Reprint those Costa del Croydon T-shirts. As you pull into East Croydon on the train, the pinky-purple colossus of Saffron Square rises in front of you, announcing your arrival into the Valley of the Crocuses. Outside the station a New Addington tram sweeps in, its doors pinging open. You ride it out past the NLA Tower, the green sweep of Lloyd Park and up into the woods, alongside the beautiful Coombe Wood Gardens, until former newsreader Nicholas Owen (for it is he who is the voice of the trams) gruffly announces, 'The next stop is Coombe Lane.' Here you alight. Dive into the trees until you emerge on the sandy heathland spine of the Addington Hills, and soon you are standing on that viewing platform presented to the town by Basil Monk in 1961.

To the north, the Shard and the Walkie-Talkie, and all those other self-important Central London confections on the horizon, look so far away, so *insignificant*. But *Croydon*, beyond these yellow gorsy slopes, postures in the foreground: muscular, strapping, vital-looking. A city in everything but name. Call it what you will: the English Alphaville, Little Manhattan, Edge City, Third City, Gotham City, Mega-City One, Croydonopolis. It is the greatest city there never was, and probably will never be.

And to prove it, the haze of urban fug lifts, and Croydon opens out before you as a living story book. Listen, and you can hear the bells of the Minster pealing, just as they have from this church since the fifteenth century. Listen more intently: are those the opening notes of *Hiawatha's Wedding Feast*? Ahead are the dazzling floodlights of Selhurst Park, where Crystal Palace might be on the verge of another 5-0 thrashing of Man United. Inside the nearby Brit School some propitious will-be is already piecing together the first notes of an album that, in a year or three, will be playing on headphones around the world.

You begin to feel light-headed. The IKEA chimneys seem to be pluming with smoke, then turn into the gothic spires of the Atmospheric Railway. Beneath the spindly Crystal Palace transmitter tower you can make out – you could swear – spectres dancing waltzes on the Beulah Spa lawn, fireworks bursting above them. As a Gatwick-bound plane emerges through the clouds, it remoulds into a Handley Page, packed with champagne and flappers, and thundering towards Le Bourget. And there, in the foreground – could it be? – surely not – a *brand-new Westfield?*

Suddenly the entire landscape is seized by a deafening, crackling, pulverising roar. It is a million times louder than the Damned at the Greyhound. You turn. And there it is: the One Croydon tower blasting off from its foundations – surging up into the ether on its cushion of flame like one of B. H. Robinson's Ladybird book illustrations. Another pummelling blast. And then another, and another. An army of skyscrapers, office blocks, car parks are pushing away the surrounding cranes like launchpad umbilicals and shooting for the Moon. Croydon is at it again. Always ahead of the game. Reaching – beyond its grasp? – for the next big thing.

Anything is possible in Croydon, after all.

ACKNOWLEDGEMENTS

I've enjoyed every minute of writing this book, and would like to give special thanks to Graham Coster of Safe Haven for having that first pint of Bass with me over which the idea was sparked, putting his trust in me to write it, and supporting me over many more pints along the way. A huge thanks also to the incredible team at the Museum of Croydon and Croydon Archives, including Mandy Smith, Rosie Vizor, Abby Pendlebury and Ruth Ilott. Croydon Libraries were indispensable too. Thank you as well to Ian Walker for giving me that tour of Croydon Airport in January 2016, which began my love affair with Croydon, and for his help with the airport chapters of the book. And thanks to Bill Lehane for lending me his copy of *Croydon Past*. I'll give it back sometime, promise.

Almost everyone I reached out to for this book was happy to get involved, which goes to show just how much Croydon is adored. Thank you to all the following, and even if you don't get a direct mention in the book, know that you helped me out in some way: Jim Collins and Ben Spicer at Addington Forge; Sean from Addington Palace; Addiscombe & Shirley Park Residents' Association; Luke Agbaimoni aka Tube Mapper; Shaun Williamson and everyone at Barrioke; Stuart Bell; Shaniqua Benjamin; Revd Canon Dr Andrew Bishop; Angela, Tony and the regulars at Bishops; Tessa Boase; Lucy Bradley; Janet Chambers; Chi-chi Nwanoku CBE at Chineke!; Georgina

CROYDONOPOLIS

Cook; Linda Connolly; Shirley McGill and the Croydon Bellringers; David Lands and Steve Thompson at Croydon CAMRA; Mark Russell at the Cronx Brewery; John Hickman, Carole Roberts and Tony Skrzypczyk at the Croydon Natural and History Society; Croydon Tourist Office (the band); Don Betson and everyone at Croydon FC; David Lavelli and Adrian Winchester at the David Lean Cinema; Steven Downes; Maddy Duxbury; Peter Fox; Sophie Marrow and Ashley Whitehouse at Foxtons; Tony Francis; Friends of Croydon Palace; John Grindrod; Catherine and Meg Gunn; Historic Croydon Airport Trust; Paula Edwards, Bill Wood, Trevor James and David Clifford at the John Whitgift Foundation; Caroline Jones; Gordon Pratt at KenEx; Vincent Lacovara; Norman Lebrecht; Lindsay Ould at Little Manhattan; Jenny Lockyer; Craig Nicholson, Alun Pemlyn and everyone at the London Canal Museum; Bishop Rosemarie Mallett; David Morgan; Katrina Navickas; Sue Perkins; Rob Plummer; Lettuce Kemp and Deborah Brown at Pollock's Toy Museum; Dave Reeves; Hilary Richardson; Ameena Rojee; Nadia Rose; Patrick Ryan; Chris Shields; Rich Simmons; Johnnie Herbert, Simon and Vida on Surrey Street Market; Andy Benham, David Edwards, Richard Perry and Ian Sutcliffe at TfL/London Trams; Paul Talling; Asifa Lahore and Shepherd's Bush at Their Majesties; David Hambleton at the Trojan Museum Trust; Turf Projects; Andrew Layzell at Waddon Leisure Centre; Alison Cousins and Mick Taylor at the Wandle Industrial Museum; David Welch; David Wickens; Deborah Klass, Joan Harry, Coral J. Reid, Jennifer Williamson, Fay Ruddock, Millicent Reid and everyone at the Windrush Generation Legacy Association; Jeremy Wooding; Robin Woods; Stuart Worden at the Brit School; and anyone else who took the time to help me out, in particular on the Facebook groups Croydon Greyhound, Croydon and Surrounding Areas History Group, and LOST Croydon (and surrounding area) forever gone . . .

Finally a huge thanks to friends and family for their support and taking an interest: Finally a huge thanks to friends and family for their support and

ACKNOWLEDGEMENTS

taking an interest: Andrew and Aislin; Laurie Ayres; Charlie and Lucinda; Dave and Jenny; James FitzGerald; Sally Foreman; Jack and Kerry; Joe and Marti; the whole team at Londonist; Diane Longman; Max Longman; Robyn Lynch; Mum and Dad; Archie Noble; Michael Ogden; Ollie and Jess; Rob and Moira; Harry Rosehill; Joanna Rutherford; Vincent and Valentina; and James Winfield. I know as soon as this book is printed I'll realise I've forgotten someone, so thanks and sorry to that person.

As soon as this book is printed I know I'll realise I've forgotten someone, so thanks and sorry to that person.

CROYDONOGRAPHY

BOOKS AND ESSAYS

The Aerodrome Hotel Croydon, replica booklet

Alsop Architects, *Third City* (Alsop, 2007)

Anderson, J. C., *A Short Chronicle Concerning the Parish of Croydon* (SR Publishers, 1970)

Andrew, Martin, *Croydon Living Memories* (Frith Book Company, 2000)

Back, Les, *So . . . fucking Croydon* (Department of Sociology, Goldsmiths College, 2003)

Baker, Eddie, *On the Run: A History of Croydon Fire Brigade* (Jeremy Mills Publishing, 2004)

Bannerman, Ronald, *A Scrapbook of Old Croydon* (Croydon Times, 1934–5)

Barnet, Christopher, *John Whitgift: Elizabeth I's Last Archbishop of Canterbury* (Whitgift Foundation, 2015)

Baron, Helen, *Trains in D. H. Lawrence's Creative Writing* (Bloomsbury Academic, 2019)

Barson, Susie, Cole, Emily and Harwood, Elain, *Tall Buildings: Aspects of their Development and Character in England* (Historic England, 2017)

Bayliss, Derek A., *Retracing the First Public Railway* (Living History Publications, 1981)

Beavis, Jim, *The Croydon Races* (Local History Publications, 1999)
Bellars, E. J., *The Canal That Failed* (1975)
Bethel, Amy, *Suburbia, Seaside and Sensation* (University of York, 2015)
Boase, Tessa, *Etta Lemon: The Woman Who Saved the Birds* (Aurum, 2021)
Boase, Tessa, *London's Lost Department Stores* (Safe Haven, 2022)
Bogle (née Nash), Joanna, Cluett, Douglas and Learmonth, Bob, *The First Croydon Airport 1915–28* (Sutton Publishing, 1977)
Bourne Society, *Local History Records: The Journal of the Bourne Society* Volume 71 (Bourne Society, 2012)
Brown, John William (compiled by), *Black's 1861 Guide to Croydon* (Adam & Charles Black, 1861)
Brownlow, Ken, *David Lean: A Biography of the Director of* Doctor Zhivago, The Bridge on the River Kwai *and* Lawrence of Arabia (Faber & Faber, 1997)
Burney, Fanny, *Stage, Screen and Sandwiches: The Remarkable Life of Kenelm Foss* (Athena Press, 2007)
Calder, Barnabas, *Raw Concrete: The Beauty of Brutalism* (William Heinemann, 2016)
Cherry, Bridget and Pevsner, Nikolaus, *The Buildings of England – London 2: South* (Yale University Press, 1983)
Chesshyre, Tom, *To Hull and Back: On Holiday in Unsung Britain* (Summersdale, 2010)
Christie, Agatha, *Death in the Clouds* (Dodd, Mead and Company, 1935)
Churchill, Winston, *Thoughts and Adventures* (Thornton Butterworth, 1932)
Clark, Sheilagh, Redsull, Robin and Thornhill, Lilian, *Conservation Areas of Croydon* (The Croydon Society, 1987)
Coe, Reginald H., *Local Buildings and Their Story* (1938–40)
Cooper, Terrence, ed., *The Wheels Used to Talk to Us: A London Tramwayman Remembers* (Tallis Pub., 1977)

Corke, Helen, *D. H. Lawrence, The Croydon Years* (University of Texas Press, 1965)

Cluett, Douglas (compiled by), *The First, the Fast and the Famous: A Cavalcade of Croydon Airport Events and Celebrities* (Hyperion Books, 1985)

Cluett, Douglas, Learmonth, Bob and Nash, Joanna, *The Great Days: Croydon Airport 1928–39* (Sutton Publishing, 1980)

Coster, Graham, *The Flying Boat That Fell to Earth: A Lost World of Air Travel and Africa* (Safe Haven, 2018)

Creighton, Sean, *Croydon's Black and Anti-Racism History 1948–79: An Introduction* (2021)

Crofts, Freeman Wills, *The 12.30 From Croydon* (Hodder & Stoughton, 1934)

The Croydon Advertiser

Croydon Guide 1969, 1973

Croydon Millenary Pageant, souvenir programme (1960)

Croydon Natural History and Scientific Society, *Croydon in the 1940s and 1950s* (Croydon Natural History and Scientific Society, 2000)

Croydon Natural History and Scientific Society, *Victorian Croydon Illustrated* (Croydon Natural History and Scientific Society, 1979)

Croydon Official Guide 1981/2, 1985/6

Croydon Official Guide Charter Centenary Year 1983

Croydon Oral History Society, *A Century of Spoken History – Talking of Croydon No. 5: Surrey Street Market* (Croydon Oral History Society, 1994)

Croydon Oral History Society, *Seventy Years of Spoken History: Talking of Croydon No. 4: Shops and Shopping* (Croydon Oral History Society, 1992)

The Croydon Times

Currie, Ian and Davison, Mark, *Surrey in the Seventies: Photographs and Memories of the 1970s* (Frosted Earth, 1995)

Deep-London Magazine

Dickson, Charles C., *Croydon Airport Remembered: An Aviation Artist Looks Back* (Hyperion Books / Sutton Libraries, 1985)

Doyle, Arthur Conan, *The Adventure of the Norwood Builder* (Strand Magazine, 1903)

Eyles, Allen and Stone, Keith, *The Cinemas of Croydon* (Keystone Publications, 1989)

Frost, Thomas, *The Old Showmen and the Old London Fairs* (Tinsley Brothers, 1875)

Gent, John, Harman, Ken and Packham, Roger (eds), *A View of Croydon: Postcards from the Past* (Croydon Natural History & Scientific Society, 2011)

Gent, John, *Croydon A Pictorial History* (Phillimore, 1991)

Gent, John, *Croydon Old and New* (Croydon Natural History & Scientific Society, 1975)

Gent, John, *Croydon Past* (Phillimore, 2002)

Gent, John, *Edwardian Croydon Illustrated* (Croydon Natural History & Scientific Society, 1981)

Green, Jeffrey, *Samuel Coleridge-Taylor: A Musical Life* (Routledge, 2011)

Grindrod, John, *Concretopia: A Journey Around the Rebuilding of Postwar Britain* (Old Street, 2013)

Grindrod, John, *Iconicon: A Journey Around the Landmark Buildings of Contemporary Britain* (Faber & Faber, 2022)

Grindrod, John, *Outskirts: Living Life on the Edge of the Green Belt* (Sceptre, 2017)

Groom, Chris, *Rockin' and Around Croydon: Rock, Folk, Blues and Jazz in and Around the Croydon Area 1960–80* (WOMBeAT Publishing, 1998)

Hadfield, Charles, *Atmospheric Railways: A Victorian Venture in Silent Speed* (David & Charles, 1967)

Harris, Oliver, *The Archbishops' Town: The Making of Medieval Croydon* (Croydon Natural History & Scientific Society, 2005)

Harley, Robert J., *Croydon Tramways: A History of Trams in the Croydon Area from 1879 to 1951* (Capital Transport, 2004)

Harris, Oliver, *Cranes, Criticism and Croydonisation: The Reshaping of Central Croydon 1937–70* (1993)

Harvie, K. G., *The Tramways of South London and Croydon 1899–1949* (London Borough of Lewisham, 1968)

Hatherley, Owen, ed., *The Alternative Guide to the London Boroughs* (Open City, 2020)

Heald, Henrietta, *Magnificent Women and their Revolutionary Machines* (Unbound, 2021)

Historic England, *The Late 20th-Century Commercial Office: Introductions to Heritage Assets* (Historic England, 2013)

Hooks, Mike, *The Archive Photographs Series: Croydon Airport* (The History Press Ltd / Chalford Publishing, 1997)

Hooks, Mike, *Croydon Airport: The Peaceful Years* (The History Press Ltd / Tempus Publishing, 2002)

Hudd, Roy, *A Fart in a Colander: The Autobiography* (Michael O'Mara, 2009)

Jackson, A. S., *Imperial Airways and the First British Airlines 1919–40* (Terence Dalton, 1995)

James, Henry, *The Letters of Henry James* Volume I (Charles Scribner's Sons, 1920)

Judd, Alan, *Inside Enemy* (Simon & Schuster UK, 2014)

Kingsley, Jason, *The Best of Judge Dredd* (Prion, 2014)

Lancaster, Brian, *'The Croydon Case': Dirty Old Town to Model Town* (Croydon Natural History and Scientific Society, 2001)

Lodge, David, *Paradise News* (Vintage, 1991)

Lovett, Vivien, *Kennards of Croydon: The Store That Entertained to Sell* (Vivien Whitchouse, 2000)

Lovett, Vivien, *Surrey Street Croydon: A Stall Story: 100 Years of Market Trading* (Frosted Earth, 1995)

Malden, H. E., *A History of the County of Surrey, Vol. 4* (London, 1912)
Marriott, Oliver, *The Property Boom* (Hamish Hamilton, 1976)
McGow, Peter, *Archives* (kindly provided by the Wandle Industrial Museum)
McKay, Sinclair, *The Secret Life of Bletchley Park: The History of the Wartime Codebreaking Centre by the Men and Women Who Were There* (Aurum, 2011)
Moore, Ald. H. Keatley and Sayers, W. C. Berwick, *Croydon and the Great War* (Libraries Committee of the Corporation of Croydon, 1920)
Morris, J., *Lecture on the geology of Croydon, in relation to the geology of the London basin and other localities* (F. Baldiston, 1875)
Nairn, Ian, *Modern Buildings in London* (London Transport, 1964)
National Trust, *Edge City Croydon* (National Trust, 2016)
O'Connor, Gerry, *The Secret Woman: A Life of Peggy Ashcroft* (Orion, 1998)
Page, William, *My Recollections of Croydon 60 Years Since* (No publisher, 1880)
Old & New Croydon Illustrated (Croydon Advertiser Group, 1979)
Parker, Eric, *Highways & Byways in Surrey* (Macmillan and Co, 1908)
Pelton, John Ollis, *Memorials of Croydon Within the Crosses* (1891)
Percy, F. H. G., *Whitgift School: A History* (Whitgift Foundation, 1991)
Phelps, Nicholas A., *On the Edge of Something Big: Edge-City Economic Development in Croydon, South London* (Liverpool University Press, 1998)
Pudney, John, *The Seven Skies: a Study of BOAC and its Forerunners Since 1919* (Putnam, 1959)
The Railway Handbook 1944–5 (Railway Publishing Co, 1943)
Rojee, Ameena, *Crocus Valley* (RRB Photobooks, 2023)
Ryde, Kenneth, *Croydon Through the Ages* (Croydon Advertiser, 1956–8)
Salter, Brian J., ed., *Retracing Canals to Croydon and Camberwell* (Living History Publications, 1986)

Sands, Revd Nigel, *Images of Sport: Crystal Palace Football Club* (The History Press, 1999)
Savage, Jon, *England's Dreaming Tapes* (Faber & Faber, 2005)
Sayers, W.C. Berwick, *Croydon and the Second World War* (Croydon Corporation, 1949)
Schüler, C. J., *The Wood that Built London: A Human History of the Great North Wood* (Sandstone Press, 2021)
Searle, Muriel V., *Down the Line to Brighton* (Bookpoint, 1986)
Shields, Chris, *The Beulah Spa 1831–56 A New Pictorial History* (chrisshieldsmusic.com, 2020)
Skinner, M. W. G., *Croydon's Railways* (Kingfisher, 1985)
Smith, Sally, *Magnificent Women and Flying Machines: The First 200 Years of British Women in the Sky* (The History Press, 2021)
Stamp, Gavin, *Britain's Lost Cities* (Aurum, 2007)
Stewart, Frances D., *Around Haunted Croydon* (AMCD Publishers, 1989)
Suburban Press
Surtees, Robert, *Jorrocks' Jaunts and Jollities* (New Sporting Magazine, 1831–4)
Thornhill, Lilian, *From Palace to Washhouse: A Study of the Old Palace, Croydon, from 1780 to 1887* (Croydon Natural History and Scientific Society, 2003)
Tyler, Kieron, *Smashing It Up: A Decade of Chaos with the Damned* (Omnibus Press, 2017)
Warwick, Alan Ross, *The Phoenix Suburb* (Blue Boar Press, 1972)
West, George, *To Penge by Canal* (Crystal Palace Foundation, 1976)
Williams, Simon, *1–2 Cut Your Hair: The Story of Johnny Moped* (Damaged Goods Books, 2022)

WEBSITES

airportofcroydon.com
architectsjournal.co.uk

CROYDONOPOLIS

bbc.co.uk/sounds

breweryhistory.com

britishnewspaperarchive.co.uk

cinematreasures.org

cpfc.co.uk

croydoncreativedirectory.com

croydonist.co.uk

curamagazine.com

dezeen.com

digitaldrama.org/little-manhattan

greatbritishlife.co.uk

guardian.com/uk

gutenberg.org

historiccroydonairport.org.uk

independent.co.uk

insidecroydon.com

jstor.org

lego.com

londonist.com

museumofcroydon.com

qantas.com

ra.co

radpresshistory.wordpress.com

railalbum.co.uk

rocknrollroutemaster.com

timeout.com

turf-projects.com

wikipedia.org

CROYDONOGRAPHY

ON SCREEN

All of Us Strangers (2023)
Airport (1934)
Are They Hostile? (2022)
Basically, Johnny Moped (2018)
Crollywood Movie Location Tour (2018)
Croydon: The High-rise and Fall (2016)
The Dark Knight Rises (2012)
The Electric Leg (1912)
Electric Transformations (1909)
The Gentlemen (2023)
Harry Hill's TV Burp
Iron Man 3 (2013)
The Late Show (1993)
Little Manhattan (2023)
The Man Who Never Made Good (1914)
The Nervous Curate (1910)
Peep Show (2003–15)
The Pirates of 1920 (1911)
Punk in London (1977)
The Return of Mr Bean (1990)
Stormzy's Glastonbury headlining set (2019)
8 Bar: The Evolution of Grime (BBC Storyville, 2023)
Unforgettable (1983)
Velvet Goldmine (1998)
Various short videos on BFI Player, Dailymotion and YouTube

MUSIC

I've listened to a lot of music to inspire and inform this book, but here are some stand-out tracks:

'Back To Black', Amy Winehouse
'Are They Hostile?', Bad Actors
'2nde Floor, Croydon', Burnin Red Ivanhoe
The Song of Hiawatha, Samuel Coleridge-Taylor
'Croydon Tourist Office', Croydon Tourist Office
'Dive', Olivia Dean
'Israelites', Desmond Dekker & The Aces
'Smash It Up', 'New Rose', 'Neat Neat Neat', The Damned
'Flagpole Sitta' (aka the theme from *Peep Show*), Harvey Danger
'Amy, Wonderful Amy', Jack Hylton
'Darling, Let's Have Another Baby', Johnny Moped
'Young Kingz Part 1 – Krept & Konan', George the Poet
Oratorio of Hope, London Mozart Players and various other artists
'They Don't Know', Kirsty MacColl
'Streets of London', Ralph McTell
'Bring Me Sunshine', Morecambe and Wise
'Saturday Gigs', Mott the Hoople
'Loudmouth', The Ramones
'Escapsim.' RAYE
'Station', 'Skwod', Nadia Rose
'Midnight Request Line', Skream
'Croydon', Captain Sensible
'Not That Deep', 'Shut Up', 'Know Me From' Stormzy

INDEX

Illustrations are highlighted in **bold**

28 Days Later, zombie film by Danny Boyle, 2
Abershawe, Jerry, 'Laughing Highwayman', 39
ABC Croydon, 191, 196
Addington Hills, 20, 50, 56, 145, 256
Addington Palace, 23, 25, **27**, 28, 29, 33
Addiscombe Place (see also East India Company), 6, 56, 58
'Adventure of the Norwood Builder, The' (see also Holmes, Sherlock, 45–6
Adele, 211
Aerodrome Hotel, 108, 119
All of Us Strangers (film), 149, 246
Allen, Sislin Fay, 176, **177**, 255
Allders (department store), 80, 82, **83**, 219, 233
Alsop, Will, 248
Amery, Colin, architectural historian, 5
Anerley Arms (see also Conan–Doyle and Sherlock Holmes), 45
Archbishops of Canterbury, 9, 10, 11, 18, 25, 26, **28**, 33, 56, 71
Armstrong, Louis, 169, 171
Arnhem Gallery, 188
Ashcroft, Peggy, 7, 8, 168–9, 190
Ashcroft Theatre, 188, 190
Atmospheric Railway, 64–5, 244, 257

Bacon, Sir Francis, 3
Bagenal, Hope, 187, 189
Balfour, Jabez Spencer, 73
Bankruptcy, council's effective, 239
Banksy, 215
Barker, Cicely Mary (see also Flower Fairies), 52–3
Barwell, Gavin, 3
Bassey, Shirley, 192
Batten, Jean, 96–7
Beano's record store, 208
Beatles, 86, 183, 191
Benga, 212, 213
Benjamin, Shaniqua (see also Poet Laureate), 50, 252
Benson, Edward White, 26, 27, 28, 29
Bentley, Derek, and Christopher Craig, 209
Bernstein, Leonard, 188
Betjeman, John, 3, 190
Beulah Spa, 54–5, **56**, 57–8, 257
Big Apple Records, 212, 245
Bishop, Andrew, Vicar of Croydon, 32
Blackadder Goes Forth, 1
Black Sheep bar, 213
Bolshoi Ballet, 165
Bonaparte Records, 197, 208
Bourne (stream), 70–1
Bowie, David, 4, 7, 192–3
Boyle, Danny, 2
Brick by Brick, 234, 235, 238, 244
Brighton, 38, 39
Brighton Road (A23), 39

273

Brit School, 210–12, 245, 246, 257
Britten, Benjamin, 189
Brooker, Charlie (see also *Black Mirror*), 144
Brown, Derren, 32
Brunel, Isambard Kingdom, 64
Bubble cars (see also Trojan cars), 153, **155**
Burton, Decimus (see also Beulah Spa), 54, 56

Caldcleugh, Alexander, slave trader, 6
Cantona, Eric, 60–1
Captain Sensible (see also Damned, the), 147, 199, 202, **203**, 209
Cartoon music venue, 207–8
Cat killings, 6, 235
Central Croydon Station, 67–8
Chaplin, Charlie, 92–4, 99, 163
Charles, Ray, 192
Chequer Inn, 16
Chesshyre, Tom, 3, 49
Chineke! Foundation and Orchestra, 186–7, 245
Christie, Agatha (see also *Death in the Clouds*), 115–6
Churchill, Winston, 101
Cinatra's, 196
Cobham, Alan, 94, 105
Coleridge-Taylor, Samuel, **7**, 8, 37, 168, 183–4, **185**, 186–7, 192, 245, 254
Conan-Doyle, Arthur (see also Holmes, Sherlock), 45, 168
Cook, Norman, 200
Coraline, novella by Neil Gaiman, 32
Corbett, Ronnie, **7**, 8, 25, 168, 209
Corps of Stewards (see also Fairfield Halls), 193
Craig, Christopher, see Bentley, Derek
Cronx Brewery, 245
Croydon: absorbed into Greater London, 179; applications for city status, 6, 130, 179, 228; cars, 128, 151–8; charcoal burning, 12, 15; Coombe Cliff Gardens, 256; Crocus Valley, 51, **52**; Croham Hurst Woods, 50; decline, 227–33; demolition, 126, 132, 171–5; department stores, 77–80, 145, 219, 228, 233; depravity, 70; derivation of name, 50; Disreputable Triangle, 70; en route to elsewhere, 10, 38, Roman times, 51; film industry, 86, 162–5; first Black woman police constable, 176, **177**; First World War, 87–9, 117; George Street, 76, 88, 152, 173, 238; ghosts, 29; Green Belt, 49; health improvement measures, 71–2; horse racing and racecourses, 19, 20, 72; hovertrain plans, 144–5; Indian community, 28; knife crime capital, 6; last hanged highwayman, 39; Little Manhattan, 136; Lloyd Park, 50, 125, 256; microcosm of Britain, 244; multi–storey car parks, 6, 9, 128, 155–6, **157**, 181; Norbury, 37; Norwood, 45, 56; Old Town, 10, 29, 70; Park Hill, racecourse, 20, park, **51**, 53, villas, 69, water tower, 243; Poet Laureate, 8, 50, 252; population, 68, 225, 249; railways and stations, 36, 40, 63, 67; redevelopment, 125–59; Riddlesdown, 50; Roundshaw, Downs, 50, estate, 249; Second World War, 118–21, 126–8, 166–7; shopping, 73–84, 145–9, 219–20, 225, 246; South Norwood, 45, 47, 65, 157, 214; Space Age, 139–41, 145; Surrey or South London? 6; theatres, 120, 161–2, 288, 190, 221, 232; Thornton Heath, 39, 154, 186, 208, 254; town hall, 72; Ultimate Rotten Borough, 6, 240; vehicle production, 40, 153–5; vertical Kew Gardens, 248; viewing platform, Addington Hills, 126, **127**, 256; Wellesley Road, 152, 247; Zeppelin raids, 88–9
Croydon Advertiser, 72, 83, 119, 128, 132, 136, 158, 169, 172, 187, 192, 199, 204, 223, 243
Croydon Airport, 81–2, 85–7, opening, 89–90, **91**, 92–123, aviatrixes, 95–101, crashes, 102–4, 121–2, new airport, 106–8, **109**, 110, Second World War, 118–22, closure, 122, last flight, 122–3, 244, preservation, 249–50, **251**

INDEX

Croydon Boxpark, 239–40
Croydon (Old) Palace (see also Old Palace School), 10, 11, **13**, 14, 21, 22, 29
Croydon Art College, 4
Croydon Basket (carriage), 40
Croydon Beekeepers' Association, 50
Croydon Boxpark, 61
Croydon Canal, 43, 44, 45, 46, 47, 243
Croydon Clocktower, 221, 222
Croydon Corporation/Council, 72, 74, 128, 155, 173, 187
Croydon Corporation Act, 130
Croydon Extra, 3
Croydon facelift, 6
Croydon FC, 251–2
Croydon Flyover, 152
Croydon Honey, 50
Croydon, John, 174
Croydon, slave ship, 6
Croydon College/School of Art/Technical College, 4, 61, 131, **157**, 174, 187
Croydon High School, 172
Croydon Minster, 29, 1867 fire, 30, **31**, 32, 52, 256
Croydon Natural History and Scientific Society, 252
Croydon Underpass, 152, **153**, 198
Croydon Water Palace, 218–9, 244
Croydonisation, 126, 128–49, 152, 154–8, 172–5, 178, 244
Crystal Palace, 20, 58, 86, 166, 186, 257
Crystal Palace FC, 59–61, 257

Damned, the, 199, 200, 202, **203**, 204, 228, 257
Davidson Road School (see also Lawrence, D. H.), 37
David Lean Cinema, 163, 201, 221, 228, 245
Davis Theatre, 120, **165**, 166, 169, 171, 188
Death in the Clouds (see also Agatha Christie), 115
Dekker, Desmond (see also 'Israelites'), 154, 245
Dickens, Charles, 22
Doctor Who, 35, 36
Doctor Zhivago, 84

Downes, Steven (see also *Inside Croydon*), 7, 237
Drew, Jane, 158–9
Driscoll, Jimmy (see also Kennards), 78–80, 96, 112
Drummond Centre, 220
Duchess of Kent, 147

East Bridge House, 174, **175**
East Croydon Station, 67, 68, 69, 142, 152, 225, 247, 256
East India Company Military Seminary (see also Addiscombe Place), 6, 56
Eden, Anthony, 91
Electric Light Orchestra (ELO), 197–8, 202
Elgar, Edward, 184–5
Elizabeth I, 18, 19, 21, 23, 31
Elizabeth II, 82, 129–30
Ellis, Havelock, 161

Fairfield Halls, 8, 145, **157**, 183, 186–8, **189**, 190–3, 195, 197, 204, 221, 228, 234, 238, 247
Faraday, Michael, 54
Farrell, Colin, 2
Fatboy Slim, see Cook, Norman
Fin, Fur and Feather Folk, 50
Fitzgerald, Ella, 171
Flower Fairies (see also Barker, Cicely Mary), 52–3
Fox, Peter, 198, 200, 208, 246

Gaiman, Neil, 32
George IV (Prince Regent), 38, 40
Get Carter (film), 158
Girls Aloud, 26
Glastonbury Festival, 214–5
Goering, Hermann, 118
Grant, Hugh, 26
Grants (department store), 36, 80, **81**, 82, 84, 228
Great North Wood, 10, 12, 56
Green Line coaches, 49
Greyhound pub and music venue, 39, 40, 43, 198–201, 203, 205, 207, 228, 257
Grim, the Collier of Croyden, 13

275

Grime, 213–5
Grindal, Edmund, 12, 14, 30, 178
Grindrod, John, 136, 149, 248

Haley, Bill, and the Comets, 169, 191
Handley Page HP-42, 108, **109, 111,** 121, 168, 257
Harding, Miss Kathleen, 174–5
Hay, Will, 97
Heath, Edward, 180, 240
Hendrix, Jimi, 195
Henry III, 11
Henry VIII, 3, 14, 21, 25, 26
Hiawatha's Wedding Feast (see also Coleridge Taylor, Samuel), 184, 257
Hill, Harry, 1
Hinkler, Bert, 94, 98
Hodgson, Roy, 60
Holly, Buddy, 169
Holmes, Sherlock (see also Conan-Doyle, Arthur), 45
Holt, Allan, borough engineer, 130, 134, 173
Home Office (see also Lunar House), 140, **141**
Hope, Bob, 25
Hudd, Roy, 166
Hyams, Harry (see also Lunar House), 139–42
Hynde, Chrissie, 202

IKEA, 214, 217–8, **219,** 257
Imperial Airways, 106–17, 121
Inside Croydon, 7, 232, 233, 237, 240
Iron Man 3, 2
'Israelites' (see also Dekker, Desmond and Trojan Records), 154

Jam, the, 199
James, Brian (see also Damned, the), 203
James, Henry, 29
Jessie J, 211
Johnny Moped, 202, 228
Johnson, Amy, 37, 97–8, **99,** 100–1, 250
Johnson, Boris, 233
Jorrocks' Jaunts and Jollities, novel by R. S. Surtees, 2

Judd, Alan, 2
Judge Dredd (Carlos Ezquerra), 178

Kenley Airfield, 119–20
Kennards (department store), 77–80, 96–7, 112, 119, 233
Kingsford Smith, Charles, 'Smithy', 94, 114

Lacovara, Vincent, 156, 248
Lambretta scooters (see also Trojan cars), 153–4
Lawrence, D. H., 37, 49
Lean, David, 3, 84, **167,** 169
Lego, 67
Lemon, Etta, 50
'Let him have it!', 209
Lewis, Jerry Lee, 191
LGBTQ+ community, 221
Lindbergh, Charles, 85–86, **87**
Liverpool, Lord, 40, 56
Lodge, David, 2
London Borough of Culture, 239, 245
London's Dullest Plaque, 4, 74
London Government Act, 179
London Symphony Orchestra, 188, 234
Lord Haw-Haw, 120
Luder, Owen (see also *Get Carter*), 157
Lunar House (see also Hyams, Harry), 140, 247–8

MacColl, Kirsty, 197, 202, 245
Macdonald, Ramsay, 91
Mad Professor, 208
Mallet, Rosemarie, Bishop of Croydon, 6, 245
Marks & Spencer, 220
Marshall, Sir James, 128, **129,** 136–7, 152, 158, 172, 178, 179
Mayday (distress call), 104, (hospital), 221
McEwan, Ian, 2
McGowan, Shane, 199–200
McLaren, Malcolm, 205
McTell, Ralph, 208–9
Melua, Katie, 211
Mitcham Lavender Water, 53–4
Mollison, Jim, 94, 100–1

INDEX

Monk, Alderman Basil, 126, 256
Monolulu, Prince, racing tipster, 3
Morecambe and Wise, 190
Moss, Kate, 3, 26
Mott the Hoople, 197
Museum of Croydon, 7

Nairn, Ian, 147, 176
National Front, 158
'Nervous Prostration', poem by Anna Wickham, 2
Nestlé (see also St George's House), 134–6, 139, 144, 158, 180, 233
Never Mind the Bollocks, Here's the Sex Pistols, 205
New Addington, 145, 158, 225, 256
Nine Lessons and Carols, 26
NLA Tower, see One Croydon
Norfolk House, 130, **131**, 134
Nwanoku, Chi-chi (see also Chineke! Foundation), 186–7

O'Connor, Des, 169
Old Palace School, 22, 23, 29, 32, 240–1
On Chesil Beach, novel by Ian McEwan, 2
One Croydon, 142, **143**, 144, 174, **175**, 178, 252, 256, 257
Orchid Ballroom, 196–7
Oval Tavern, 8, 252
Overground, London, 44
Owen, Nicholas, 256

Page, Hilary Fisher (see also Lego), 67
Parsons, Tony, 204
Pearce, Guy, 2
Pearce, Jonathan, 60
Peep Show (David Mitchell and Robert Webb), 1, 222–4
Penry, John, 17, 18
Pepys, Samuel, 56
Perkins, Sue, 1, 4
Pevsner, Niklaus, 143, 147, 178
Philips, Eliza, 50
Picasso, 222
Pickles the dog (see also World Cup), 58
Pitt, William, the Younger, 40
Previn, André, 189–90

Proby, PJ, 196
Punk rock, 198–205
Purley Way (A23), 36, 37, 106, 120–1, 152, 217, 218
Purley Way Lido, 112, **113**, 218, **220**

Queen Victoria, 58, 68

Ramones, the, 199
Rattle, Simon, 188
RAYE, 211, 246
Redding, Otis, 197
Reeves Corner (and House of Reeves), 227, 230, **231**
Reid, Jamie (see also *Suburban Press*), 173, 205, 255
Ribbentrop, Joachim von, German ambassador, **119**
Riesco, Raymond, 188, 232
Rinse FM, 212, 213
Riots, 227, 230–2
Ritchie, Ian, 241
Robinson, B. H. (Ladybird books illustrator), 139, 257
Rogers, Richard, 171, 247
RSPB, foundation of, 50
Rojee, Ameena, 50
Rose, Nadia, 209, 256
Rossi, Francis (see also Status Quo), 128
Royal Festival Hall, 187–8
Royal Philharmonic Orchestra, 166, 188
Royal School of Church Music, 27

St George's House (see also Nestlé and Greyhound pub), 134, **135**, 144, 198, 235, 237, **239**
St George's Walk, 144, 146, 148, 207, **229**, 235, 237
Saffron Square, 249, 256
Sainsbury's, first in suburbia, 76, **77**, first self-service, 145–6, biggest in world, 147, Whitgift Centre store closure, 240
Samuda brothers (Jacob and Joseph), 64, 66
Sanderstead, 246
Scabies, Rat (see also Damned, the, 202, 203

Scott, Andrew (see also *All of Us Strangers*), 149
Scott, Giles Gilbert (see also Croydon Minster), 30
Seifert, Richard (see also One Croydon), 142–4, 177
Selfridge, Gordon, 112
Selhurst Park, 59–61
Selsdon Man, 180, 240
Selsdon Park Hotel, 49, 180, **181**, 240
Seven Hills of Croydon, 156
Sewell, Brian, 5
Sex Pistols, 205
Sheldon, Gilbert, 30
Silver Wing flights, 108–9, 121
Simmons, Matthew, 60
Siouxsie and the Banshees, 201
Skream, 212, 213
Southgate, Gareth, 61
Stanley, Bob, 209
Star pub, 192, 195
Starmer, Keir, 197
Status Quo, 128, 193
Stiff Records, 204
Stokowski, Leopold, 188
Stormzy, 213–4, **215**, 244, 254
Stravinsky, Igor, 189
Strummer, Joe, 197
Suburban Press (see also Reid, Jamie), 4, 148, 173
Surrey Iron Railway, 41, 42, 44, 243
Surrey Street Market, 4, 70, 209, 213, 252
Surtees, Robert Smith, 2

Taberner House, 132, **133**, 174
Talking Heads, 199–200
Talling, Paul (*Derelict London*), 228, 230
The 12.30 from Croydon, 108–9
The Song of Hiawatha (see also Coleridge Taylor, Samuel), 184
Their Majesties LGBTQ+ club, 221, 252, **253**
Thornton, Big M ama, 192
To Hull and Back, 3
Top Rank club, see Cinatra's
Tower of Light, 5
Townshend, Pete, 188
Tram crash, 6, 236–7

Trams, 73–5, 82–4, 145, 225, **226**, 227, 236–7, 246, 250–1, 256
Trinity School of John Whitgift, 32, 172, **173**
Trojan cars (see also bubble cars, Monk, Alderman Basil), 153–4, **155**
Trojan Records (see also Dekker, Desmond), 154
Turpin, Dick, 39

Underground music venue, 205

Vanian, Dave (see also Damned, the), **203**
Veteran Car Run, 37

Walton, Izaak, 16
Wandle, River and Valley, 13, 41, 42, 51, 252
Warehouse Theatre, 221, 232
Waters, Muddy, 192
West Croydon Station, 44, 63–4, 186
Westfield Croydon, 233, 237, 240, 257
Whitehouse, Mary, 196
Whitgift, John, 13, 14, 16, **17**, 22, 32, 129; temper, 16; anti–Puritanism, 17; closeness to Elizabeth I, 19, 21, 22; possible gay relationship, 22; tomb, 29, 30, **31**; statue, 16, 72
Whitgift Almshouses, 15, 16, **18**, 19, 31, 173, 241
Whitgift Centre, 9, **11**, **32**, 146–7, **148**, Forum pub, 148, 149, 156, 172–3, 209, 233, 240, 241, 245, 246, 254
Whitgift Foundation, 31, 32, 172, 234, 240, 241
Whitgift School, 9, 32
Wickham, Anna, 2
Williams, Kenneth, 1
Williams, Rowan, 18
Windrush generation, 176, 254, **255**
Winehouse, Amy, 211–12
Wood, Wilfred, 27
Woolworths, 76
World Cup (see also Pickles the dog), 58, 61
Wrestling, 190, 234

Zaha, Wilfred, 254